Global Perspectives on Health Geography

Series editor

Valorie Crooks, Department of Geography, Simon Fraser University, Burnaby, BC, Canada

Global Perspectives on Health Geography showcases cutting-edge health geography research that addresses pressing, contemporary aspects of the health-place interface. The bi-directional influence between health and place has been acknowledged for centuries, and understanding traditional and contemporary aspects of this connection is at the core of the discipline of health geography. Health geographers, for example, have: shown the complex ways in which places influence and directly impact our health; documented how and why we seek specific spaces to improve our wellbeing; and revealed how policies and practices across multiple scales affect health care delivery and receipt.

The series publishes a comprehensive portfolio of monographs and edited volumes that document the latest research in this important discipline. Proposals are accepted across a broad and ever-developing swath of topics as diverse as the discipline of health geography itself, including transnational health mobilities, experiential accounts of health and wellbeing, global-local health policies and practices, mHealth, environmental health (in)equity, theoretical approaches, and emerging spatial technologies as they relate to health and health services. Volumes in this series draw forth new methods, ways of thinking, and approaches to examining spatial and place-based aspects of health and health care across scales. They also weave together connections between health geography and other health and social science disciplines, and in doing so highlight the importance of spatial thinking.

Dr. Valorie Crooks (Simon Fraser University, crooks@sfu.ca) is the Series Editor of Global Perspectives on Health Geography. An author/editor questionnaire and book proposal form can be obtained from Publishing Editor Zachary Romano (zachary.romano@springer.com).

More information about this series at http://www.springer.com/series/15801

Raymond Asare Tutu • Janice Desire Busingye

Migration, Social Capital, and Health

Insights from Ghana and Uganda

Springer

Raymond Asare Tutu
Associate Professor of Geography
Delaware State University
Dover, DE, USA

Janice Desire Busingye
Senior Lecturer, Deputy Vice Chancellor
Kampala International University
Kampala, Uganda

ISSN 2522-8005 ISSN 2522-8013 (electronic)
Global Perspectives on Health Geography
ISBN 978-3-030-24692-1 ISBN 978-3-030-24693-8 (eBook)
https://doi.org/10.1007/978-3-030-24693-8

This Springer imprint is published by the registered company Springer Nature Switzerland AG
The registered company address is: Gewerbestrasse 11, 6330 Cham, Switzerland

Preface

Geographic mobility has gained global currency and has become integral to the discussion on sustainable development. We use a mixed-methods approach on the internal migration dynamics regarding disease ecologies of informality and the interactions between social capital, lifestyles, health literacy, and health outcomes in the context of informal settlements in two developing countries—Ghana and Uganda. Through the prism of the concepts of place and scale, the book demonstrates the myriad of ways by which place or context directly and indirectly influences migrants' health knowledge, literacy, and outcomes in poor urban slums.

The healthy immigrant hypothesis suggests that migrants upon arrival are more likely to have better health status compared to their host populations due to positive migration selectivity. That notwithstanding, the health advantage enjoyed by migrants erodes with time and becomes identical to the health status of the host population mainly due to the process of acculturation. In the context of international migration, this puzzle is amplified by the developed–developing countries migration stream, that is, the fact that most of the immigrants to Western countries depart from countries experiencing relatively higher morbidity and mortality rates. Nevertheless, the good health status of new immigrants has been explained by the assertion that if migrants are selected from the higher socioeconomic hierarchy of the sending countries, it is more likely that such migrants will have better health care, less exposure to negative ecological factors, and the ability to afford a more nutritional diet. While a number of studies have used a variety of health variables such as life expectancy, and noncommunicable disease outcomes to compare newly arrived migrants and the host population, the literature has consistently upheld the healthy immigrant hypothesis and its link to acculturation. The processes of acculturation, precipitated by increasing duration of stay, including changes in lifestyle and behaviors as well as exposure to contextual factors have been identified as responsible for the erosion of the health advantage enjoyed by migrants.

Behavioral patterns among immigrants that may result in positive or negative health outcomes are influenced by social and contextual factors such as material deprivation, perceived stress, social capital, and perceived discrimination. The decline in health is made worse when the adoption of the host lifestyle is coupled with the experience of emotional distress stemming from family separation, for example. Additionally, the literature demonstrates that migration to cities in the developing world may increase the risk of cardiovascular disease because of the relative increase in physical inactivity and poor eating habits. Apart from the suggestions that acculturation is the cause of health outcomes, that is, either as a risk or protective factor, acculturation may also be a result of the individuals' response to risks and protective factors. This may in turn explain other pathways by which acculturation impacts health. It has been theorized and empirically found that acculturation influences health behavior through the adoption of coping strategies to personal discrimination or a perceived sense of discrimination, inadequate social capital, and vulnerability to adverse health behavior as well as changes in norms, attitudes, and values.

Despite a substantial amount of work on migration, social capital, and health outcomes in the international migration context, there is a paucity of analysis in the internal migration context that integrates all these variables into a holistic approach and focuses on health literacy among the urban poor in informal settlements. The complexities in operationalizing and developing measures for acculturation, social capital, and health literacy require sophisticated research methodologies that enhance triangulation. This is essential in understanding the multifaceted linkages between social capital, acculturation, and health (health literacy and outcomes) in places of deprivation. We employed a mixed-methods approach in exploring these linkages. We adopted quantitative methods (e.g., surveys) and qualitative methods including participatory research methods such as conceptual mapping (mental models), participatory hazard mapping, and body health mapping.

The authors' field work experience and results of the study are presented in this book. Chapter 1 provides an overview of internal migration into urban slums of both countries. We discuss the major streams of migration in the countries and demonstrate the major hubs and final destinations of migrants. We briefly describe the informal settlements chosen as study sites. This is followed by a presentation of the general methodology as well as a description of the demographic characteristics of research participants. Chapter 2 focuses on a brief general literature on the connections between migration, social capital, and health. In Chap. 3, the patterns of diseases are discussed. We begin with a discussion on the disease patterns in both countries. We proceed to discuss communicable and noncommunicable diseases reported to the immediate clinics in the study sites to illustrate double burden of disease among the urban poor. Self-reported disease patterns among migrants in the communities are examined using participatory body health mapping. Environmental risks to health are focused in Chap. 4. Environmental contaminants are tested and migrant's perceptions and knowledge on food safety given the presence of environmental contagions. A description of the environmental risks to health in these informal settlements with emphasis on exact types of environmental exposures and contaminants are provided. Chapters 5 and 6 discuss the varied conceptions of health literacy and examine the health literacy of migrants on environmental risks to health, predominant infectious diseases in the informal settlements, and health literacy on noncommunicable diseases. Chapter 7 investigates the association between social capital, acculturation, lifestyles, and perceptions on health outcomes while Chapter 8 assesses the association between social capital and health literacy.

Dover, DE, USA — Raymond Asare Tutu
Kampala, Uganda — Janice Desire Busingye

Acknowledgement

We are grateful to the entire research team in both Uganda and Ghana. The fieldwork team leaders in Uganda were Mr. Jordan Y.L. Byekwaso and Francis Garuzooka. Byekwaso is a Part-Time Assistant Lecturer in the Department of Adult and Community Education, School of Distance and Lifelong Learning, College of Education and External Studies, Makerere University, Kampala. He holds a Master and Bachelor of Adult and Community Education from Makerere University. Besides the 9 years of engagement in NGO work, Jordan also is passionate about research engagement and has been involved in a number of research projects in the fields of adult education, community development/empowerment, education for sustainability, social entrepreneurship, HIV/AIDS, poverty, gender, food security, lifelong learning, higher education, education in general, water quality and affordability among others. Further, Jordan is committed to pursue a PhD (by research) in the area of lifelong learning and social innovations (considering quality, access, and impact).

Francis is an Assistant Lecturer in the Department of Adult and Community Education, Faculty of Special Needs and Rehabilitation, Kyambogo University, Uganda. He holds a Master of Adult and Community Education from Makerere University and a Bachelor of Adult and Community Education of Kyambogo University. He has previously worked as a Programs Director at National Adult Education Association of Uganda and a Trainer of Trainers in "*start and manage micro-enterprises*" at International Labor Organization in Uganda. He is also currently participating in writing of a trainer's manual for Community Development Officers on sexual reproductive health and HIV/AIDS with Ministry of Gender, Labour and Social Development (MoGLSD). His research interests include family planning education, reproductive health, vocational education and training using participatory appraisal, and learning research methods.

The fieldwork team leaders in Ghana were Ms. Doris Ottie-Boakye and Ms. Harriet Sarkodie. Ms. Sarpong is a doctoral candidate at the Regional Institute for Population Studies (RIPS), University of Ghana. Prior to her enrollment into the PhD program at RIPS, she was a Fellow at INDEPTH Network, Ghana, a demographic surveillance system organization. Ms. Harriet Sarkodie holds a bachelor's in psychology from the University of Ghana. She has led a couple of social science research teams and worked in both study sites in Ghana.

Our sincere gratitude to all the field research assistants including Bernice Adda, University of Ghana, Kwadwo Owusu Jr. and Richard Owusu Amankwah of the Kwame Nkrumah University of Science and Technology. Others are Paul Obeng, Ruth Ampomah, Akwasi Amanin Asare, Enyonan Senaya, Stella Oppong, and Victoria Degbee. Ebiasah Sampson Odoom, who is also the assistant to the Assemblyman of Ngleshie, Jamestown, Mr. Joseph Addo, Paul Golightly, and Sanjor Tunde all have our utmost appreciation.

We appreciate the valuable time rendered to the project by Dr. Atiku Abubakar, Lecturer, Kampala International University, who assisted in the collection of the data on chemical contamination, and Dr. Muhammad Kibuuka, Lecturer in College of Economics and Management, who assisted in the analysis of the Ugandan quantitative data.

Special gratitude to the health workers and local clinics for the data. We are very thankful to all the local authorities and research participants for their support and participation in our research activities. Finally, we acknowledge Delaware State University and Kampala International University for partially funding some of the activities culminating into this book.

Contents

About the Authors

Raymond Asare Tutu, Ph.D. is an Associate Professor of Geography in the College of Humanities, Education, and Social Sciences at Delaware State University, United States. He has a Doctor of Philosophy degree in Geography from the Pennsylvania State University and a Master of Philosophy degree in Population Studies from the Regional Institute for Population Studies, University of Ghana. His research interests and scholarly publication are in the subfields of population health, global migration, human–environment geographies including urban environments with emphasis on informality and marginality, as well as water access and health ecology. His scholarly peer-reviewed publications are in top-tier journals such as *GeoJournal, Journal of Population Research, Tropical Medicine and Health, Journal of Infections and Public Health, African Geographical Review, American Journal of Tropical Medicine and Hygiene, Habitat International,* and *Journal of Housing and the Built Environment*. He has been a reviewer for scientific peer-reviewed journals including *The Social Science Journal, Rural Sociology, Journal of International Migration and Integration, African Review of Economics and Finance, International Development Planning Review, Journal of Rural Studies, Journal of Asian and African Studies*, as well as *PLOS ONE*. Courses he has taught are Global Societies, World Regional Geography, Social Change, Population Analysis, Technology and Society, and Introduction to Sociology. He is currently a board member of the Africa Specialty Group of the American Association of Geographers. Dr. Tutu is also a member of the Health and Medical Geography Specialty Group of the American Association of Geographers.

Janice Desire Busingye, Ph.D. is a Senior Lecturer and seasoned scholar in the College of Education, Open and Distance Learning and the Deputy Vice Chancellor, at Kampala International University. She holds a Doctor of Philosophy Degree in Education and a Master of Education (Adult Education) from the University of KwaZulu-Natal, South Africa. Her research and scholarly interests focus on adult education and community development in general and on the subfields of educational systems, literacy, and health literacy in particular, livelihoods and epistemology. She has previously carried out research and published peer reviewed works in journals such as the *Africa Education Review, International Social Science Journal, GeoJournal, Journal of Population Research, Tropical Medicine and Health, Journal of Infections and Public Health*, among others. She currently teaches a module on Gender and Development; previously, she has taught Policy Studies, Political Economy, Introduction to Critical Situation Analysis, and Human Relations for Community Educators. She is a member of Association of Professional Adult Educators, Uganda, and American Association of Geographers.

Chapter 1
Introduction

Abstract This chapter introduces the subject matter of this book. We begin by presenting an overview of internal migration into slum settlements in Ghana and Uganda. The historical antecedents—pre- and postcolonial—that influenced and continue to impact internal migration are discussed. The economic developmental patterns across space in both countries which have triggered diverse streams of migration such as rural to urban, urban to urban, urban to rural, and rural to rural are deliberated. The chapter further presents other variables that have influenced out-migration among the poor and vulnerable including conflicts, political instability, and environmental concerns. Furthermore, a detailed description of the study areas in both Uganda and Ghana is provided. Apart from the geographic locations, we show which informal settlements are designated officially as residential places and the ones that are not, as well as the ramifications for environmental service provision. We show a gist of the extent of deprivation, which is highlighted in other chapters. A subsection on methodology provides the overall data and methods information for the primary data collected. Information is provided on sampling strategy and methods of data collection such as survey, biological and chemical contaminant testing, body health mapping, participatory hazard mapping, and conceptual mapping.

Keywords Internal migration · Slums · Informal settlements · Old Fadama · James Town · Kinawakata · Namuwongo · Uganda · Ghana

Overview of Internal Migration into Ghana's Urban Slums

Population mobility is an age-old occurrence and has been with us since the beginning of time. In this section, we discuss the evolution of Ghana's internal migration, its trends, and dynamics over time with emphasis on the movements into urban slums. Migration has been central to the livelihoods of both rural and urban residents in Ghana since the formation of the nation. All the major streams of migration characterize Ghana's internal migration—rural–urban, rural–rural, urban–rural, and

R. A. Tutu, J. D. Busingye, *Migration, Social Capital, and Health*,
Global Perspectives on Health Geography,
https://doi.org/10.1007/978-3-030-24693-8_1

urban–urban. In the 1960s, Ghana's migration was predominantly internal migration with people moving from different regions of the country into others for safety reasons to avoid being casualties of conflicts, new arable lands, and to search for lands to resettle families (Wyllie 1977; Boahen 1975; Caldwell 1968). Regarding migration within the administrative regions of the country, the major stream has been north-south (e.g., Hashim 2007; Kwankye et al. 2007). While earlier forms of the north-south migration in Ghana were attributed to a quest for fertile land for farming, a large number of people were found to have left their homes in the north to flee ethnic conflicts, particularly in the mid-1990s (Black et al. 2006). Migration due to the search for arable lands to farm was largely seasonal in nature and primarily influenced by the variations in the ecological zones of the country; the northern half is usually dry for most of the year, while the southern part experience relatively longer periods of rain during the year (Abu et al. 2014). More recent data show that the propensity of an individual to internally migrate in Ghana is a function of a coupled individual pull factors like age and education, and community-scale push factors such as availability of social amenities (Ackah and Medvedev 2010). Specifically, while younger people have a higher probability to migrate, residents of communities with better access to water and environmental sanitation have a lower propensity to migrate (Ackah and Medvedev 2010).

The labor migration to the southern half of the country met the labor needs predominantly in the agriculture and mining sectors of the economy. In the agriculture sector, apart from the contribution of migrant labor to the growth of food crops in the "food basket" of Ghana, migrants from the north were significantly involved in the development of cocoa production, the major cash crop of Ghana (Abdul-Korah 2007). During the colonial era, labor was moved involuntarily into the mining and agricultural sectors in the southern half of the country to aid the economic pursuits of the colonialists (Abdul-Korah 2007). The migration trend continued after Ghana gained political independence with voluntary migrants moving mainly into Brong-Ahafo, and Ashanti regions to work in the farms and mines, as well as Greater Accra region to pursue other forms of manual labor. Economic structural adjustment programs and policies pushed by the Bretton Woods institutions and adopted by the Ghana government, which involved the elimination of subsidies on healthcare and agricultural fertilizers, adversely impacted rural communities and the northern part of the country thereby aggravating the north-south migration trend (Abu et al. 2014).

Government policies that favor urban areas over rural ones have been argued to be responsible for rural–urban migration. "Urban-bias" policies such as protection of industries and suppression of food prices led to low rural incomes and shift from agricultural labor thereby triggering rural out-migration (Anarfi et al. 2003). While the major migration stream within the country is rural–urban—approximately 25% of migratory moves—and attributed to family union, search for better environmental conditions, and better economic opportunities (Ghana Statistical Service 2008), the persistent north-south migration has been attributed to the spatial variation in the developmental patterns (Anarfi et al. 2003). Scholars have argued that colonial and postcolonial economic, social, and environmental policies favored the southern half of the country at the expense of the north thereby making the north the labor stock feeding the cash crop, mining, and manual cosmopolitan industries of the south (see Awumbila and Momsen 1995; Abdul-Korah 2007). This has resulted in relatively better social and economic opportunities in the southern half thereby being a major pull factor for economic migrants from the north. Anarfi et al. (2003) assert that the biased regional developmental patterns have resulted in "three geographic entities" in the country namely the coastal zone, the middle zone, and the northern savanna. The coastal zone, which has the administrative capital, is the most urbanized and industrialized followed by the middle zone. Consequently, the middle and coastal zone (southern half) have become the major destination for northern migrants due to the perceived higher wages and fertile lands they have to offer.

Earlier forms of the north-south migration were dominated by males who moved to the agricultural and mining areas for new livelihoods (Anarfi et al. 2003). This trend has changed tremendously with sizeable number of females migrating to the south largely for economic reasons (Kwankye et al. 2009; Awumbila and Ardayfio-Schandorf 2008). Several studies have examined the migration of young females from the north to the major cities like Accra and Kumasi in the south of the country to work in the markets (Awumbila and Ardayfio-Schandorf 2008; Anarfi and Kwankye 2005; Anarfi et al. 2003). These young women independently migrate to the cities to work as head porters, also known as *kayayei*, often carrying goods on their heads for a fee. Their services are used by shoppers who need their goods cart from one place to the other. Although that has become major source of livelihood for the young women, some view this trade as a temporary measure based on socio-cultural demands (Awumbila et al. 2008). For example, some of the young females migrate to become *kayayei* to earn money to be able to afford stuffs needed for marriage (Awumbila and Ardayfio-Schandorf 2008). Such items will be evidence that the female is of age and can be married thereby enhancing her chances of being a responsible bride.

Associated with the north-south migration was the creation of several ethnic enclaves in the urban areas of the southern half of the country. Of notice is the formation of the *Zongos*[1] (Harvey and Brand 1974). In a study of the spatial allocation of migrants in Accra, the administrative capital of Ghana, Harvey and Brand 1974 found that the Zongos and other poorly developed areas served as the residential areas of less skilled and poor migrants with less urban experiences from the north of the country. Such areas act as "bridgeheaders" for northern migrants who are more familiar with rural lifestyle to adjust to the new urban life. Proximity to markets, which are the main work places of the migrants, have led to the creation of informal settlements around such markets for the poor migrants (Tutu 2013). Such settlements provide cheap and affordable accommodation for the migrants as well as a sense of belonging especially among those who reside in places dominated by

[1] In Ghana, Zongo refers to a community or neighborhood with substandard environmental conditions specifically created and inhabited by migrants from the northern parts of the country. It is characterized by inadequate social amenities such as sanitation facilities, pipe water, and electricity, as well as ramshackle houses.

people of their ethnicity (Xu 2011; Tschakert and Tutu 2010). Therefore, migration into urban slums in Ghana is both for economic advantage as well as social satisfaction. Research has shown that poor migrants prefer these communities due to the perceived economic opportunities in the informal sector including jobs and business engagements deemed profitable (Tschakert and Tutu 2010). The relatively low rent cost and low cost of living is considered, by migrants, an opportunity to save part of their income.

Overview of Internal Migration into Uganda's Slums

Migration in Uganda has been a major part of history determined by conflicts between different states and during the process of colonization, especially in the western part of the country (Mukwaya, et al. 2012). The colonial era experienced significant levels of internal migration, which involved laborers moving from north-west to the south-west to work in sugar cane plantations (Lyons 1996) and other on private coffee plantations (Rutabajuka 1989). Although rural–rural migration was encouraged by the colonial government to depopulate the south western part of Uganda, the trend has not been enough to influence immigration patterns. After colonization and as industrialization became a significant player in the economy, a host of occasional and seasonal casual laborers started leaving the rural areas to the urban centers.

After independence, given the political instability that characterized Ugandan politics, there was a significant change in the migration patterns of the population (Potts 1997). During the tumultuous years of the 1970s and 1980s, the migration pattern was reversed from rural–urban to urban–rural movement, especially in the major towns, like Kampala. By 2002/2003, at least 10% of the household heads reported to have migrated in the previous 5 years (World Bank 2006). The highest rates of in-migration occurred in Kalangala District, on the shores of the Lake Victoria, followed by Kampala and Wakiso districts, respectively (Mukwaya et al. 2012). While the predominant perception is that many people move from rural to urban areas, in some parts of Uganda, specifically, the northern and western regions, the reverse was true. Most of the immigrants into Kampala tended to come from the surrounding central region. Studies have shown that the residents who were born in Kampala are less than half of the total population in Kampala and have remained constant over the past years (Nyakaana, et al. 2007).

The development of slum dwellings can be traced back to the 1600s, when Kampala was established as a Capital city of Uganda, giving it a dual urban structure, separating the local town from the municipality (UN HABITAT 2006). In the planning of the municipality, the only occupants of the city that were considered were the Whites and Indians, excluding the huge African population that occupied most of the areas in the city. Kampala back then, in the 1970s, was referred to as the "city of seven hills," while the low-lying areas that were occupied by the Ugandans were forgotten, hence the beginning of the expansion of vast areas of land into slums.

Kampala is currently the second fastest-growing city in the East African region, divided into five administrative divisions, which collectively cover a total of 189 km^2, of which 169 km^2 are covered by land and 20 km^2 of water (Ministry of Works and Transport 2009). While the size of the land is not expanding, the population of Kampala has been growing steadily. For instance, in 1970, Kampala had a population of 330,700, which increased to 774,221 in 1991, and by 2014 was estimated at 1,516,210 (Uganda Bureau of Statistics (UBOS) 2014). It is estimated that by 2020, the greater Kampala Metropolitan Area (GKMA) will have a population of 5,000,000 people. The current annual population growth rate of the urban area of Kampala is 3.9% (above the national rate of 3.3%) (KCCA 2012). The rate of growth in the urban center has been above or near 4% for 30 years. As the population grows, slums also increase because 60% of the population in Kampala occupies slum dwellings (Mukwaya et al. 2012). Such housing conditions have affected the housing conditions of slum dwellers, where most of them live in temporary structures, garages, and small rooms. Recent studies have indicated that people who stay in slum areas lack basic social services (KCCA 2012).

Description of Study Sites

The two study sites in Ghana are Old Fadama and James Town in the city of Accra, and the sites for Uganda are Nawumongo and Kinawataka in the city of Kampala. We describe their geographic locations in the cities and their exogenous characteristics.

Old Fadama

This neighborhood is considered a squatter settlement and located in the heart of the city of Accra. Old Fadama is bordered on the northwest by the Korle Lagoon and vast canals to the south and east (see Fig. 1.1). The space is not designated for residential purpose although it served as a temporary shelter or place of abode for migrants fleeing ethnic conflicts from the northern parts of Ghana in the 1980s and 1990s (Tutu 2014; Tschakert and Tutu 2010). Consequently, none of the residents has a title to the land they occupy (Grant 2006). Despite the lack of title to the land, the space functions as residential and center of trade with commercial activities such as food markets (e.g., yam and tomatoes), small scale enterprises (food vendors and hairdressers), and vendors markets (Grant 2006); a paragon of slum cities in the developing world. The official population of the slum is unknown although it was estimated to be about 80,000 in 2010 (Housing the Masses 2010). In more recent times, there have been calls for the eviction of residents for several reasons including (1) the beautification needs of the city; (2) the pursuance of an ecological project called the Korle Lagoon Restoration Project; and (3) the perception among

Fig. 1.1 Map of Old Fadama (Tutu 2013)

Accra residents that the settlement is a haven for criminals. This perception led to the settlement being nicknamed "Sodom and Gomorrah," two Biblical cities that were destroyed by God because of their evil deeds (Tutu 2014; Grant 2006). The illegal status of residents coupled with contentions over the rightful owners of the space has prevented the expansion of basic infrastructural facilities to residents. Thus, sanitation facilities, places of convenience, and garbage collection are left for residents to provide; this has resulted in enormous environmental risks to health in the neighborhood. Some of the environmental and health stressor residents face range from acute inadequate latrines and swamping of rooms during rains to struggling with mosquitoes. Due to poor layout in the community, it is one of those slums where someone's kitchen may be another's parlor. Figure 1.2 illustrates the general exogenous characteristics of the settlement.

James Town

James Town is a neighborhood located in the Ashiedu Keteku sub-Metropolitan area in Accra (Fig. 1.3). It is one of the two communities that make-up Ga-Mashie, also called Old Accra, one of the traditional areas of the Ga people. In fact, it is known to be the first place that the indigenous people settled and, therefore, the oldest settlement in the city. Although it is officially known as a slum because it meets

Fig. 1.2 Housing structures and environment in Old Fadama (authors' fieldwork)

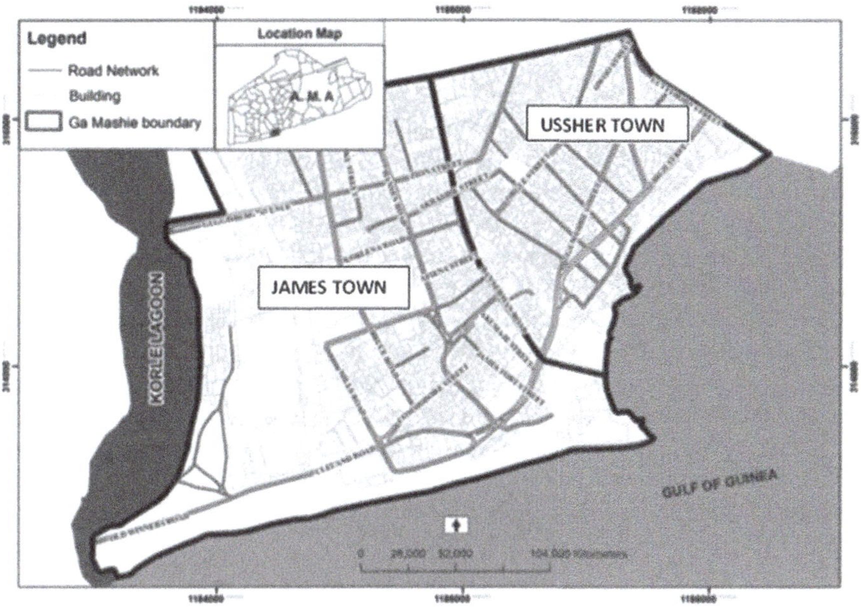

Fig. 1.3 Map of James Town, Accra (Tutu et al. 2017)

all the cardinal characteristics of a slum, it has been designated a matured one and tenure secured (AMA and UN-HABITAT 2011). The Ghana 2010 population census estimated the population of James Town to be 15,508. It is densely populated with a higher than urban average of fertility rate and household size (Tutu et al. 2017). Regarding planning, it has good road network and community layouts. James Town has attracted many migrants due to the relatively lower accommodation cost and closeness to both the Central Business District of the city as well as proximity

Fig. 1.4 Pictures of some drains and immediate surroundings of James Town (authors' fieldwork)

to some local markets where migrants ply their trade. So, while it is predominantly a traditional neighborhood, there is visible cosmopolitanism on display due to migrant ethnic enclaves formed. An obvious example is the creation of a warehouse that provides sleeping and dwelling place for migrants from Bawku, a town in the Upper East Region of the country. Thus, it serves that ethnic group and so it has been named Bawku warehouse (Schraven et al. 2013). Although the community receive social amenities services due to their status as legitimate settlement, it still lacks adequate social infrastructures thereby predisposing residents to varied health risks. James Town is among the five neighborhoods in the city having lowest housing quality and is associated with a higher risk for negative health outcomes (Weeks et al. 2012). While the community may be argued to be well laid out with tarred roads, there are glaring features of environmental risk to health (see Fig. 1.4 for pictures of some drains in this study community).

Namuwongo

Namuwongo was is located in Makindye Division of Kampala District (Fig. 1.5) and derives its name from a Luganda word "*Amasigukuludde,*" which means to exhume. It covers 85.6 acres of land, of which 30% is owned by the Municipality and 70% is privately owned.

The major source of water in Namuwongo is water taps, which are both communal and individual. However, some of the people get water from wells and springs, which they tend to rely on a lot because water is free. All water from these sources mentioned here is not safe (ACTOGETHER 2014). Sanitation in Makindye

Fig. 1.5 Map of Namuwongo's study site, Kampala (source: Google maps)

Fig. 1.6 Open well in Namuwongo (authors' fieldwork)

division is characterized by individual toilets (2910), shared toilets (10), communal toilets (5), and public toilets (19), and all these are pit latrines (ACTOGETHER 2014). In addition to the toilets, residents in informal settlements have reported to be using open defecation (30%) and the bucket system, sometimes called 'the flying toilet' (20.6%). This means that slightly more than 50% of the population have no access to and do not use toilets. Given that water is fetched from wells and springs, it means that there is high contamination of water from open sources (see Figs. 1.6 and 1.7). Garbage collection is also a challenge in these areas, given that the city council only collects garbage once a week, yet there are no designated spots for col-

Fig. 1.7 Pictures of open spring water and semi-protected spring water in Nawumongo

lecting garbage in the community. It is a common occurrence to find garbage dumped anywhere in the community to avoid paying the council, which leads to contamination too. The most common diseases that affect Namuwongo in order of priority were reported to be malaria, cough, ring worms, and diarrhea (ACTOGETHER 2014).

Kinawataka

Kinawataka is in Nakawa Division of Kampala and is the biggest settlement in the division (Fig. 1.8). It derives its name from roads that were not tarred and became muddy whenever it rained. The slum changed its name from Mbuya during the regional naming and demarcation of political divisions in the 1980s (ACTOGETHER 2017). It covers 157 acres of land of which 40% of the land is owned by the municipality, and the church and Buganda Kingdom own 30% each. It has a total of population of 80,000 people living in 4000 households. Ninety-six percent of these live in the settlement and 4% come from outside to operate small scale businesses. Most of the structures in Kinawataka are permanent (83%) and are mainly used for residential purposes. The Kinawataka slum has a rail crossing in the slum and will be prone to evictions when development on this railway line starts (ACTOGETHER 2014).

The major source of water in Kinawataka is mainly from individual taps, community taps, and open springs (see Figs. 1.9 and 1.10). Most of the water is known

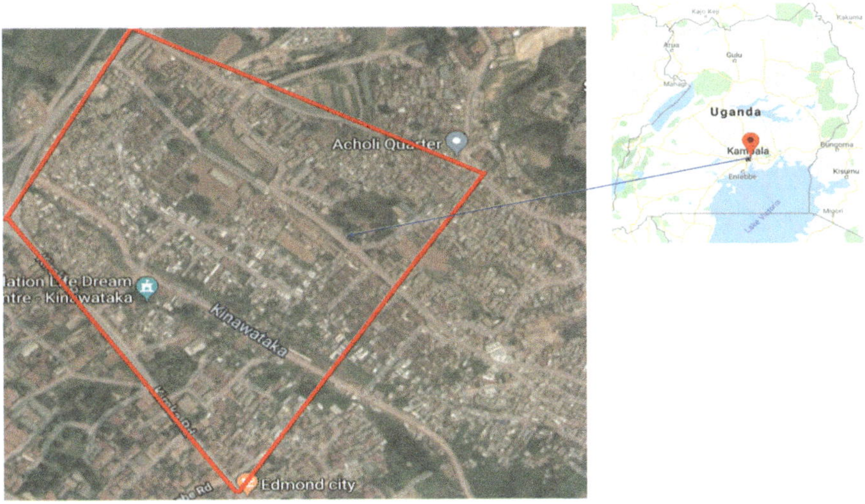

Fig. 1.8 Map of Kinawataka's study site, Kampala (source: Google maps)

Fig. 1.9 Pictures of open spring water in Kinawataka (authors' fieldwork)

to be unsafe and must be boiled before consumption. In the whole division, there are only 16 toilets, servicing a population of 80,000 people. Garbage collection is usually handled by individual collectors; however, it is common to find it thrown in ditches and drainage channels. The most common diseases that affect residents here are malaria, typhoid, HIV/AIDS, and diarrhea (ACTOGETHER 2017).

Fig. 1.10 Pictures of tap water in Kinawataka (authors' fieldwork)

Methodology

This subsection provides the general data and methods information for the study resulting in this book. Specific details are further provided within each chapter, especially regarding content of methods of data collection and analyses. Here, we provide information on sampling strategy (sample size and sampling process) for each of the methods used in the data collection process. The activities undertaken for data collection include survey questionnaire, biological and chemical contaminant testing, body health mapping, participatory hazard mapping, and conceptual mapping (also known as mental models).

For the survey, the sampling technique adapted was dependent on the nature and structure of the settlement. We segmented the community into zones and each enumerator was assigned a specific zone. The zone assigned was further divided into four. On sampling, if we did not have access to the official census enumeration maps, random sampling was adopted as follows: Enumerators and research assistants were provided a dice. They stood at the entry point of the community—the beginning points of the subdivisions—and began the sampling activity. The dice was supposed to be cast for direction, skip patterns, and selection of household. The enumerator throws the first dice to indicate the direction he/she will go. The next dice was thrown to indicate the number of houses to be skipped. The third dice was thrown only when needed—that is, when the housing structure to selected had multiple households living in it. Depending on the size of the zone, the expected outcome of the dice for skip pattern was limited. That is, it ranged from 3 to 6. Obviously, the direction was limited to four. So, enumerators rethrew the dice whenever a number above 4 appeared. Concerning the selection of household for the survey in the event of multiple households, the number for the dice was limited to the number of housing units in the housing structure. Enumerators asked to ascertain the number of households and not simply use the number of rooms in a compound house. The research assistants counted from left to right given the number of households and assigned numbers accordingly. The survey was administered by trained research assistants.

To avoid ambiguities, we clearly defined household and migrant. Our operational definition of a household is as defined by the Ghana Statistical Service for censuses and the Demographic and Health Surveys. The same definition was adapted in Uganda. A household is defined as a person or a group of individuals, who live together in the same house or compound, share the same house-keeping engagements, and are provided for as one unit. While the general notion of a household will imply a man, his wife, children, and some other relations or a house help who may be living with them and sharing the same house-keeping arrangement, it is significant to bring to mind that members of a household may not be necessarily blood related or related by marriage. This is because nonrelations like house-helps are part of a household. Conversely, not all relations living in a house and sharing the same house-keeping arrangement in a house or compound are members of the same household. This is because, for instance, two siblings living in the same house with their spouses and off-springs may or may not form distinct households contingent on their catering activities. On migration, we adapted the conventional definition of migration. That is, movement of people across space with the intention of establishing permanent or semi-permanent residence at the destination. So, individuals who have moved from any part of the country into our study sites with the intention of staying permanently or semi-permanently and has been there for not less than 6 months could be included in our study.

Training was an essential component of the data collection. While the team leaders for all sites and research assistants were experienced in field research, authors and the entire team had a rigorous training program. The content of the training consisted of behavior as an enumerator or research assistant on the field—how to approach people, informed consent, language of interview, patience and tactfulness on the field, and dealing with difficult cases; predata collection training—thought concepts and definitions of major topics; pretesting instruments among colleagues, and correction and modification of instruments; and receipt of final documents and materials for the field work—informed consent, final instruments, enumeration areas, and zone specification for each enumerator and groups of research assistants, and other materials such as pens, pencils, erasers, markers, adhesives, sticky notes, flip charts, recorders, cameras, and so on. Although several contacts have been made with community leaders prior to the beginning of the entire project, on the day of the initial activity, these opinion leaders and/or their assistants were contacted on reaching the community for homage to be paid to them in line with traditions and norms of the communities.

Research participants used the language they were most comfortable speaking. Each question on the survey questionnaire and guidelines for the body health mapping, conceptual mapping, and participatory hazard mapping was translated from English into the predominant common languages spoken in the communities by professional and native speakers. Through the training activities, we made sure of uniformity of intended meaning to all participants. Major phrases and wording of questionnaire items and guideline were reviewed continually to prevent vagueness and meanings misplaced in translation.

In both Namuwongo and Kinawataka, using satellite imagery from Google maps, we subdivided the communities into zones for sampling purposes. Namuwongo was divided into five zones, and 30 households were randomly selected from each zone. Kinawataka was divided into ten zones. We randomly selected 15 households from each of the zones. For James Town, the Ghana Statistical Service has divided the community into Enumeration Areas (EAs) for census and survey administration. James Town has 24 EAs. From a previous household listing exercise, 150 households were randomly sampled from each of the EAs in Ngleshie, James Town. However, for Old Fadama, we used aerial view to demarcate the zones. We had a total of ten zones. Fifteen households were subsequently randomly selected (as previously described) from each of the zones.

Participatory hazard mapping was used to understand migrant's perception through depiction of dangerous and safe places in the community for migrants' health. We draw on Keith and Brophy (2004) and Tschakert (2009) application of hazard mapping in the contexts of asbestos-exposed workers in Canada and artisanal gold mining in Ghana, respectively. As a tool, participatory hazard mapping is intrinsically efficacious and valid due to its participatory nature that involves participants who have close knowledge of their environment (Keith and Brophy 2004). Additionally, it enables the process of "consensual validation"—where contributors of the activity reach agreement that the information presented was precise in accordance with their shared-lived experience. The activities were segmented by sex in all four communities of the study. The number of participants per study site was 10–12 resident migrants drawn from across the length and breadth of the communities. This was to ensure that as much as possible health hazards across the community were fully captured. These activities were held in community centers or place of worship, which we rented. These were places that were known to participants and/or convenient to access. While recruitment was done within the households of participants, informed consent was delivered upon arrival at the venue for the activity. Individual consent was provided and not a general one although the essence of the activity was repeated at the beginning of the activity itself. Details of the content of the guideline as well as the activity are discussed together with the finding in Chap. 4.

Body health mapping was used to understand self-report diseases in the communities. Body mapping, a participatory research method, is a health survey tool that uses graphic images, colored markers, and sticky notes to obtain and document participant responses (Keith 2003; Keith et al. 2002). While the method was first used to understand work-related health issues linked to injuries and ergonomics, it has been applied in the collection of diverse health data (Keith and Brophy 2004). Between 30 and 40 people participated in this activity per study site, which lasted between 1.5 and 2 h. Participants were recruited from their houses and informed about the venue of the activity. Informed consent was sought at the venue from each individual participant before the activity began. No names of participants or personal identifiers were used on the body maps during activities. Activities were not segmented by sex but differentiated by color codes. These activities took place in rented or hired community centers or places of worship. These places were known to participants and/or convenient to access. See Chap. 3 for full discussion of the process.

Conceptual mapping (mental model) was used to understand (1) how migrants and medical personnel think about fecal contamination, and (2) social capital formation and its impact on acculturation and lifestyles, which subsequently affects health outcomes. Conceptual maps are mental representations of actual or theorized circumstances in the form of individual maps (Morgan et al. 2002). Conceptual map interviews consist of obtaining one's beliefs about an issue (e.g., hazard) as expressed in the individual's own terms. Morgan et al. (2002, p. 20) stress that such "interview protocol is shaped by the influence diagram, so that it covers potentially relevant topics. It allows the expression of both correct and incorrect beliefs and ensures that the respondents' intent is clear to the interviewer." The method is premised on philosophical stance such as the truism that although people might lack a complete understanding of a particular risk phenomenon, they still possess some beliefs about that phenomenon. Such beliefs stem from what they may have heard about the phenomenon, which might prompt them about related issues and inform some mental associations and linkages. All these will be assembled through a mental model after which decisions are made. We draw on Morgan et al. (2002) conceptions and Tschakert and Singha (2007) application of mental models in the context of artisanal gold mining for our study. Six migrants and two health professionals per community participated in this activity. Like the survey, the activity was held in the houses of respondents are at the location that was most convenient for them. With the health professional, it was either in the health facilities where they worked or at a venue of their choosing.

Demographic Characteristics of Study Respondents

In this subsection, the sociodemographic background information of our study population is presented. We begin with the two-study sites in Ghana. Subsequently, the demographic characteristics of the respondents from the two sites in Uganda are described.

James Town and Old Fadama, Ghana

Eighty-three percent of the respondents from James Town were females, whereas females constituted 68% of the study population in Old Fadama (Table 1.1). While 10% of the study population in James Town reported to have no education, 31% in

Table 1.1 Background information of study respondents, James Town and Old Fadama

	James Town			Old Fadama		
Variables	*N*	%	Means	*N*	%	Means
Sex						
Female	125	83.30		102	68.00	
Male	25	16.70		48	32.00	
Education						
No education	15	10.00		47	31.30	
Primary	16	10.70		28	18.70	
Junior high school	68	45.30		49	32.70	
Senior high school	30	20.00				
Senior high and higher				26	17.30	
Tertiary	21	14.00				
Ethnicity						
Akan	95	63.30		27	18.00	
Ewe	10	6.70		8	5.30	
Mole-Dagbani	15	10.00		84	56.00	
Guan				10	6.70	
Gurma				10	6.70	
Grusi				2	1.30	
Other	30	20.00		9	6.00	
Religious affiliation						
No religion	4	2.70				
Christian	136	90.70		58	38.70	
Muslim	10	6.70		92	61.30	
Religiosity						
Not at all	2	1.30		6	4.00	
Occasionally	15	10.00				
Less often	17	11.30		8	5.30	
Often	30	20.00		26	17.30	
Very often	48	32.00		58	38.70	
Always	38	25.30		52	34.70	
Marital status						
Single	38	25.30		43	28.70	
Cohabiting	12	8.00		21	14.00	
Separated/divorced/widowed	22	14.70		9	6.00	
Married	78	52.00		77	51.30	
Employment status						
Unemployed	41	27.30		13	8.70	
Employed	109	72.70		137	91.30	
Occupation						
Professional/clerical	14	9.30		6	4.00	
Sale/service	73	48.70		79	52.70	
Manual	23	15.30				

(continued)

Table 1.1 (continued)

	James Town			Old Fadama		
Variables	*N*	%	Means	*N*	%	Means
Skilled manual				25	16.70	
Unskilled manual				27	18.00	
Unemployed/students	41	26.70		13	8.70	
Age			40.05			29.46
Age at migration			19.81			18.67
Duration of stay (years)			16.18			8.71
Household size			4.50			3.52
Total	*150*	*100*		*150*	*100*	

Old Fadama indicated that they have had no formal education. Twenty and 14% reported to have senior high school and tertiary education, respectively, in James Town compared to 17% in Old Fadama with senior high school or higher education. This reflects the general notion that, while both places are slums because they lack major infrastructural facilities, Old Fadama is more of an abode for the less educated poor.

Regarding ethnicity, while majority of respondents in James Town identify as Akans (63%), most (56%) of the study population in Old Fadama indicated that they are of the Mole-Dagbani ethnic group. This confirms the discussions in the literature that most of the inhabitants of Old Fadama are migrants from the northern parts of the country. Ewes constituted about 7 and 5% of the respondents in James Town and Old Fadama, respectively. "Other" category refers to all other ethnic groups originating from Ghana other than Akan, Ga-Dangme, Ewe, Guan, Gurma, Mole-Dagbani, Grusi, and Mande, in consonance with the categorization by the Ghana Statistical Service.

Concerning religious affiliation, most of the respondents in James Town (90%) profess to be of the Christian faith, while majority of the participants in Old Fadama (61%) reported to be of the Islamic faith. This also reflects other writings on these communities. Old Fadama is known to be a predominantly Muslim community due to its status as the major destination for migrants from the Muslim majority northern parts of Ghana. On religiosity—the extent to which people practice and adhere to the tenets of their faith—the data suggest perception of high adherence. About 77% of the respondents in James Town often, very often, and always practice and adhere to the teaching of their faith compared to 91% in Old Fadama.

Fifty-two and 51% of the study population were currently married in James Town and Old Fadama, respectively. While 25% reported to be single in James Town, 29% indicated their marital status as single in Old Fadama. In James Town, 8% of the study population reported to be cohabiting, whereas 14% indicated they were cohabiting in Old Fadama.

Regarding employment and work status, 73% of the respondents in James Town stated that they were gainfully employed compared to 91% of migrant respondents in Old Fadama. The predominant occupations were sales/service and manual labor—skilled and unskilled. Sales and service occupations include, but not limited to, petty trading, shop keeping, bar tending, and food vending. While skilled manual labor comprise hairdressing, barbering, aluminum fabricator, mechanic, and masonry, unskilled manual encompass, but not limited to, metal scrap dealing, head porterage, and laborer (landscaping). In James Town, 49% of the respondents said they were in sales/service compared to 53% in Old Fadama. Fifteen percent of respondents in James Town were engaged in manual labor compared to 35% of the respondents in Old Fadama. The large percentage of manual labor in Old Fadama is attributed to the large number of our respondents who are engaged in metal and electronic scrap dealership around the community and head porterage.

The mean age of respondents in James Town is 40 years, while the mean age of migrants interviewed in Old Fadama is 29 years. The mean age at migration is almost even at about 20 and 19 years in James Town and Old Fadama, respectively. The average duration of stay in James Town is 16 years among our respondents, while the mean length of stay in Old Fadama is 9 years. The lower length of stay in Old Fadama may confirm the evidence in the literature that the community is one of the most transient migrant communities in the city of Accra. Household size among our respondents is about five and four in James Town and Old Fadama, respectively.

Kinawataka and Nawumongo, Uganda

Seventy-three percent of the respondents in this study from Namuwongo are females compared to 74% from Kinawataka (Table 1.2). Respondents who did not have any formal education were 16% for Namuwongo and 18.5% for Kinawataka, representing a minimal difference. However, while Kinawataka had 47% of the respondents indicating primary school attainment, in Namuwongo 43% had attained ordinary level education.

Thirty-one and 25.3%, which is more than half of the study population in Namuwongo slum reported to be of Acholi and Luo ethnicity, respectively, coming from the northern region of Uganda, while Kinawataka has a high presence of Bagisu, Basamya, Banyole, and Bagwere (46.4%), all from neighboring districts in Eastern Uganda. This confirms popular belief that most slum dwellers migrated to these slums because of the civil war in Northern Uganda and more recently, the mudslides around the Mt. Elgon area in Eastern Uganda that affected many families.

Concerning religious affiliation, most of the respondents in Namuwongo slum reported to be Christians (91.3%), while 82.8% in Kinawataka reported to be of the Islamic faith. This is contrary to the writings about Ugandan populations that put Christians as the majority in most communities in Uganda. In terms of practicing their faith, both groups of respondents, in both slums, do not demonstrate a high religious practice culture. Those who reported to practice always were 8.7% for Namuwongo and 20.5% for Kinawataka.

Forty-one percent reported to be married in Kinawataka compared to 30.7% in Namuwongo. In both slums, the statistics show that there is a high number of people who are cohabiting, which is interpreted as marriage in most Ugandan societies.

Table 1.2 Demographic characteristics of respondents in Namuwongo and Kinawataka

	Namuwongo			Kinawataka		
Variables	*N*	%	Means	*N*	%	Means
Sex						
Female	110	73.30		111	73.50	
Male	40	26.70		40	26.50	
Education						
No education	24	16.00		28	18.50	
Primary school	26	17.30		71	47.00	
O level	63	42.00		36	22.50	
A level	12	8.00		13	8.60	
Tertiary	05	3.30		5	3.30	
Ethnicity						
Acholi	47	31.30		1	0.70	
Luo	38	25.30		4	2.60	
Iteso	4	2.70		11	7.30	
Langi	2	1.30		4	2.60	
Munyakitara	10	6.70		12	7.90	
Muganda	20	13.30		29	19.20	
Bagisu and Basamya	15	10.00		29	19.20	
Bagwere and Banyole	7	4.70		41	27.20	
Basoga	3	2.00		6	4.00	
Others	4	2.70		14	9.30	
Religious affiliation						
Traditionalist	4	2.70		23	15.20	
Christian	137	91.30		3	2.00	
Muslim	9	6.00		125	82.80	
Religiosity						
Not at all	7	4.70		1	0.70	
Occasionally	28	18.70		23	15.20	
Less often	55	36.70		38	25.20	
Often	35	23.30		27	17.90	
Very often	11	7.30		28	18.50	
Always	13	8.70		31	20.50	
Marital status						
Single	35	23.30		33	21.90	
Cohabiting	44	29.30		37	24.50	
Divorced	6	4.00		1	0.70	
Separated	18	12.00		16	10.60	
Married	46	30.70		62	41.10	
Employment status						
Unemployed	75	50.00		79	52.30	
Employed	75	50.00		72	47.70	
Occupation						

(continued)

Table 1.2 (continued)

Variables	Namuwongo N	%	Means	Kinawataka N	%	Means
Casual worker	7	4.70		12	7.90	
Technician	3	2.00		7	4.60	
Armed forces/security	5	3.30		1	0.70	
Foods and beverage seller	27	18.00		11	7.30	
Crafts and related trades	6	4.00		5	3.30	
Building and construction	4	2.70		4	2.60	
Service and sales workers	3	2.00		13	8.60	
Agric, forestry, and fisheries	3	2.00		3	2.00	
Market vendor	13	8.70		15	9.90	
Professionals	4	2.70		1	0.70	
Others and not working	28	18.70		79	52.30	
Don't know	47	31.3				
Age			31.69			30.70
Age at migration			15.82			16.27
Duration of stay (years)			11.49			7.48
Household size			5.088			4.29
Total	*150*	*100*		*150*	*100*	

Regarding employment, 50% of the respondents in Namuwongo reported to be gainfully employed, while 50% were unemployed. In Kinawataka, the number of the unemployed (52.3%) was higher than that of those who are employed. Most of those who were employed were working in the informal sector with 18% in Namuwongo working as food venders, 8.7% as market vendors, and 4.7% as casual workers. In Kinawataka, 8.6% reported to be food vendors, 9.9% worked as market vendors (head porters), and 7.9% were casual workers.

The mean age of respondents in Namuwongo is 31.6 years, while the mean age of respondents in Kinawataka is 30.7 years. The mean age at migration is almost the same in both slums, with Namuwongo at 15.8 years while Kinawataka is at 16.3 years. The average duration of stay in Namuwongo is 11 years while for Kinawataka is 7 years. This could indicate the differences in the more recent circumstances that led to their migration to Kampala. Household size among the respondents was five for Namuwongo and four in Kinawataka.

References

Abdul-Korah G (2007) 'Where is not home?': Dagaaba migrants in the Brong Ahafo region, 1980 to the present. Afr Aff 106:71–94

Abu M, Codjoe SNA, Sward J (2014) Climate change and internal migration intentions in the forest-savannah transition zone of Ghana. Popul Environ 35(4):341–364

Accra Metropolitan Assembly and UN-Habitat (2011) Participatory slum upgrading and prevention millennium city of Accra, Ghana. Resource document. https://www.yumpu.com/en/document/view/11527678/participatory-slum-upgrading-and-prevention-the-city-of-accra

Ackah C, Medvedev D (2010) Internal migration in Ghana: determinants and welfare impacts. The World Bank. http://econ.worldbank.org

ACTOGETHER (2014) Kampala City slum profile 2014. Kampala City report. ACTOGETHER. UNPUBLISHED

ACTOGETHER (2017) Kinawataka Nakawa Enumeration report 2017. Kampala City report. ACTOGETHER. UNPUBLISHED

Anarfi J, Kwankye S (2005) The costs and benefits of children's independent migration from Northern to Southern Ghana. Paper presented at the international conference on childhoods: children and youth in emerging and transforming societies, Oslo, Norway, 29 June–3 July 2005

Anarfi J, Kwankye S, Ababio OM, Tiemoko R (2003) Migration from and to Ghana: a background paper, University of Sussex, DRC on Migration, Globalisation and Poverty

Awumbila M, Ardayfio-Schandorf E (2008) Gendered poverty, migration and livelihood strategies of female porters in Accra, Ghana. Norwegian J Geogr 62(3):171–179

Awumbila M, Momsen JH (1995) Gender and the environment: women's, time use as a measure of environmental change. Glob Environ Chang 5(5):337–346

Awumbila M, Manuh T, Quartey P, Tagoe CA, Bosiakoh TA (2008) Migration country paper (Ghana). Centre for Migration Studies, University of Ghana, Legon

Black R et al (2006) Migration and development in Africa: an overview. African Migration and Development Series No. 1, Southern African Migration Project (SAMP)

Boahen AA (1975) Ghana: evolution and change in the nineteenth and twentieth century. Longman, London

Caldwell JC (1968) Determinants of rural-urban migration in Ghana. Popul Stud 22(3):361–377

Ghana Statistical Service (2008) Ghana living standards survey report of the fifth round (GLSS 5). Ghana Statistical Service, Accra

Grant R (2006) Out of place? Global citizens in local spaces: a study of the informal settlements in the Korle Lagoon environs in Accra, Ghana. In: Urban forum, vol 17, no 1. Springer, Netherlands, pp 1–24

Harvey ME, Brand RR (1974) The spatial allocation of migrants in Accra, Ghana. Geogr Rev 64(1):1–30

Hashim I (2007) Independent child migration and education in Ghana. Dev Change 38(5):911–931

Housing the Masses (2010) Community-led enumeration of Old Fadama Community, Accra—Ghana. In: Final report to People's Dialogue on Human Settlements. Accra, Housing the Masses

Kampala Capital City Authority (KCCA) (2012) Promoting green urban development in African cities: urban environmental profile of Kampala, Uganda. https://www.kcca.go.ug/uDocs/Kampala_Urb_Env_Profile.pdf

Keith M, Brophy J, Kirby P, Rosskam E. Barefoot Research: A Work Security Manual for Workers. Geneva, Switzerland: International Labor Organization, 2002.

Keith M (2003) Workplace health and safety mapping: the why and how of body mapping. Occup Health Rev 102:31–33

Keith MM, Brophy JT (2004) Participatory mapping of occupational hazards and disease among asbestos-exposed workers from a foundry and insulation complex in Canada. Int J Occup Environ Health 10(2):144–153

Kwankye SO, Anarfi JK, Addoquaye Tagoe C, Castaldo A (2007) Coping strategies of independent child migrants from Northern Ghana to Southern Cities. Development Research Centre on Migration, Globalisation and Poverty Working Paper Series, University of Sussex

Kwankye SO, Anarfi JK, Tagoe CA, Castaldo A (2009) Independent north-south child migration in Ghana: the decision-making process. Development Research Centre on Migration, Globalisation and Poverty, University of Sussex Working Paper T-29

Lyons M (1996) 'Foreign bodies': the history of labour migration as a threat to public health in Uganda. In: Nunget P, Asiwaju AI (eds) African boundaries: barriers, conduits and opportunities. Pinter, London

Ministry of Works & Transport (2009) National transport master plan including a transport master plan for the greater Kampala Metropolitan Area (NTMP/GKMA). https://works.go.ug/wp-content/uploads/2015/08/National-Transport-Master-Plan-2008-2023.pdf

Morgan MG, Fischhoff B, Bostrom A, Atman CJ (2002) Risk communication: a mental models approach. Cambridge University Press, Cambridge

Mukwaya P, Bamutaze Y, Mugarura S, Benson T (2012) Rural-urban transformation in Uganda. J Afr Dev 14(2):169–194

Nyakaana JB, Sengendo H, Lwasa S (2007) Population, urban development, and the environment in Uganda: the case of Kampala City and its environs. Department of Geography, Makerere University, Mimeo

Potts D (1997) Urban lives: adopting new strategies and adapting rural links. In: Rakodi C (ed) The urban challenge in Africa: growth and management of its large cities. United Nations University Press, New York, pp 447–494

Rutabajuka S (1989) The conditions of migrant labour in Masaka District 1900–1962: the case of coffee shamba laborers. Centre for Basic Researc (CBR) Working Paper no. 7. CBR, Kampala

Schraven B, Hinz A, Renaud P, Rumke C, Schommers A, Sikorski A (2013) Youth poverty in Accra: managing urban livelihoods in informal apprenticeships. Resource document. International Labour Office: Employment Sector, Skills and Employability Department, German Development Institute—Geneva. https://www.die-gdi.de/uploads/media/Employment_Report_21_2013.pdf

Tschakert P (2009) Digging deep for justice: a radical re-imagination of the artisanal gold mining sector in Ghana. Antipode 41(4):706–740

Tschakert P, Singha K (2007) Contaminated identities: mercury and marginalization in Ghana's artisanal mining sector. Geoforum 38(6):1304–1321

Tschakert P, Tutu R (2010) Solastalgia: environmentally induced distress and migration among Africa's poor due to climate change. In: Environment, forced migration and social vulnerability. Springer, Berlin, pp 57–69

Tutu RA (2013) Self-rated resilience among young migrants in old Fadama, Accra, Ghana. GeoJournal 78(4):709–725

Tutu RA (2014) Dilemmatic experiences of young migrants in Accra, Ghana: the merciless hands of pirate urbanization and landlordism. J Hous Built Environ 29(4):637–656

Tutu RA, Boateng JK, Busingye JD, Ameyaw E (2017) Asymmetry in an uneven place: migrants' lifestyles, social capital, and self-rated health status in James Town, Accra. GeoJournal 82(5):907–921

Uganda Bureau of Statistics (UBOS) (2014) National census report 2014. UBOS, Kampala. https://www.ubos.org/wpcontent/uploads/publications/03_20182014_National_Census_Main_Report.pdf

UN HABITAT (2006) Cities without slums sub-regional programme for Eastern and Southern Africa. Situation analysis of informal settlements in Kampala. Kivulu (Kagugube) and Kinawataka (Mbuya 1) Parishes. http://www.rrojasdatabank.info/kampala.pdf

Weeks JR, Getis A, Stow DA, Hill AG, Rain D, Engstrom R et al (2012) Connecting the dots between health, poverty and place in Accra, Ghana. Ann Assoc Am Geogr 102(5):932–941

World Bank (2006) Uganda Poverty and vulnerability assessment report. Report no. 36996-UG Poverty Reduction Management, Southern Africa, Africa Region. World Bank, Washington, DC

Wyllie RW (1977) Kponoe and Tado stool - problem in interpretation of Anlo Migration Tradition (Ghana-Togo). Anthropos 72(1–2):119–128

Xu L (2011) Inter-metropolitan migration of the newly landed immigrants in Canada: 1991–1996 and 1996–2001. Geojournal 76(5):501–524

Chapter 2
Migration, Social Capital, and Health

Abstract In this chapter, the interconnections between migration, social capital, and health is reviewed. The purpose is to show the inextricable link between these concepts through the lenses of theoretical and empirical literature on the subject. The chapter presents a short description on drivers of migration trends and the "globalization of migration" phenomenon. Through the prism of migration systems theory, we discuss the macrostructural and microstructural factors that shape migration. We zoom in on microstructural factors such as social networks and their contribution to both facilitating population movement and aiding in adaptation at the destination. Furthermore, the role of social networks in enhancing social capital notwithstanding, arrival of migrants at the destinations predisposes them to varying stressors. The stressors are more pronounced in the urban periphery, which host most poor migrants. Irrespective of the poor conditions in the urban periphery, the literature shows that migrants' health statuses upon arrival are usually better than their hosts, especially with respect to chronic diseases; a situation referred to as the "epidemiological paradox." The chapter concludes with a discussion on how social capital mediates migration and health outcomes.

Keywords Migration · Social capital · Health · Epidemiological paradox · Social networks · Migrants' adaptation · Urban migrants · Migration systems

Human migration takes different forms and have had diverse patterns over time. Recent migration trends are attributed to both old and emerging factors such as economic shocks, political instability, cultural diversity, environmental change, and violent conflicts. Irrespective of these varying but connected reasons, Castles et al. (2013) argue that there are certain common tendencies of contemporary patterns of migration. Firstly, more and more countries are being affected by population movements at the same time. The "globalization of migration," as they referred to it, has most migrant receiving countries having entrants from wider range of social and cultural upbringings. Secondly, population movements have increased in volume and in stock across all regions. This is buttressed by evidence from the United Nation's study on the trends in international migration (United Nations 2015).

R. A. Tutu, J. D. Busingye, *Migration, Social Capital, and Health*,
Global Perspectives on Health Geography,
https://doi.org/10.1007/978-3-030-24693-8_2

Thirdly, migration is differentiated. That is, most countries have multiple types of migration—labor migration, refugees, permanent status, or settlement. Additionally, these multiple types reveal themselves in diverse ways sometime all at once and perpetuated by migratory chains. The fourth tendency has been called "feminization of migration." While earlier forms of migration were predominantly involved males and gender variables less analyzed—labor migration and refugee movements, after the 1960s, females are no longer only perceived and classified in family reunion migratory patterns but have been captured in labor migration and other streams as well. Next, migration is being politicized. This is happening at different scales of governance. The need for migration policy to incorporate global governance tenets will encourage cooperation between sending, transit, and receiving countries.

The latter trend justifies the theorization of migration as a system. A migration system's theory proposes that population movement comes about because of prior connection between sending and receiving countries. The connections are premised on factors such as on family and social networks, trade, colonization, cultural links, and global political influences (Castles et al. 2013). More broadly, these factors may be categorized into macrostructural factors and microstructural factors. Macrostructural factors entail issues that pertain to the role of international affairs and the involvement of states or governments in the organization and facilitation of population movements across space (Hollifield 2004). These include practices, laws, and international relationships, which are meant to facilitate or minimize migration, between given sending and receiving countries. Microstructural factors on the other hand are the social networks that are cultivated by migrants at the destination to enable them to settle and survive at the destination. Such networks are the hub of cultural and social capitals. Cultural capital in the forms of travel information about the destination, knowledge of migration to preferred countries, organization of travel, as well as adaptation in the destination facilitates the initiation of population movement and enables its sustenance. Social capital is exemplified in the important resources that individuals and groups obtain through family, household members, friends, community bonds, and reciprocity in financial and social issues (Boyd 1989). Social capital through the networks at the destination expedites the processes of settling down and community construction. This is done through the development of social infrastructure including ethnic associations, grocery shops, religious meeting places, and professional networking.

On arrival at the destination, migrants are exposed to many stressors that affect them negatively. Recent population increase in cities and urban periphery, because of internal migration and increasing fertility rates (Chattopadhyay et al. 2006), has put a strain on social amenities; thus, environmental health problems including infectious disease epidemics have been exacerbated (Fobil et al. 2012). Continual exposure to undesirable conditions such as residential struggles, perceived and actual acts of bias and prejudice, overcrowding, high number of people per household, unsanitary practices, as well as poor living conditions at the destination have been intimated as being the main explanations for racial health inequalities among migrants (Harris et al. 2006). Therefore, although the urban space may have been known for improved health services access, improved social services, and better electricity and water access, central government's inability to expand services to match the increasing population in the cities in most developing countries has led to poor environmental conditions and the associated undesirable health outcomes among migrants (Jankowska et al. 2015). Fundamentally, rural–urban migrants living in poor communities and on the fringes of urban societies have a higher likelihood of acquiring and or dying from a communicable disease.

These unpleasant conditions notwithstanding, evidence from research on migrants' health statuses show that immigrants are usually healthier than the host populations of the destination countries especially regarding noncommunicable diseases and mental ailments. Apart from the verification from individual migrants, the same is true for migrant households. Data also show that members of migrant households are healthier than host population household members. This situation is referred to as the "epidemiological paradox" (Lu 2008). The paradox has been attributed to, among other factors, migration selectivity with respect to age and health. According to Lu, at the heart of migration decision-making is health concerns. Younger people are more likely to migrate because they are more likely to be healthier and with a better coping stamina and inner tenacity to withstand problems associated with migration. Additionally, in the case of labor migration, which may require more physical strength and high endurance levels, older people and less healthy individuals have lower probability of migrating.

The "epidemiological paradox" over time dovetails into the "healthy migrant hypothesis"—a situation where the health of migrants deteriorate over time although newly arrived immigrants from poor countries of origin have better health status than individuals of the host countries (Palloni and Morenoff 2001). The corrosion of the health of the newly arrived migrant begins at the onset of the adoption of unhealthy behaviors and lifestyles of the dominant culture including the hustle and bustle in the city with its concomitant lifestyle (Torun et al. 2002). This is further compounded by the emotional and psychological difficulty the migrant may suffer due to family separation and the likely loneliness that may ensue at the destination (Lu 2010). Migration to cities in the Global South has been linked to rising probability of cardiovascular diseases due to poor lifestyles at the destination—decreased physical activity and adoption of poor eating habits. Torun et al. (2002) intimate that migration to the city is related to the consumption of food that are high in saturated fat and processed sugar while not enough legumes and vegetables.

Microstructural factors mediate migration and health outcomes. Social networks across space play diverse roles, which impact migrant's health positively and negatively across scale. Over time, social capital constructs such as civic participation, reciprocity, and trust have been found to be associated with better mental health, better self-rated health status, lower cardiovascular deaths, and lower suicide rates (Helliwell 2003; Poortinga 2006; Fone et al. 2007; Kawachi et al. 1997). Whereas at the individual scale social capital is related with ability to cope with stress, illnesses, and maintenance of good health (Tutu 2013; Uchino et al. 1996), social capital manifested in, for example, social cohesion is linked with lower extent of violent crimes, reduction in the negative impacts of neighborhood deprivation, and improved mental health (McKenzie et al. 2002; Kennedy et al. 1998).

One of the seminal contributions to understanding the relationship between social capital and health is by Szreter and Woolcock (2004). They argued the need to differentiate three forms of social capital: bonding, bridging, and linking. This was to enhance our understanding specific pathways through which social relations impact health outcome. Their contribution emerged due to the inadequacies of three perspectives on social capital and its implications for public health. These are social support perspective, inequality proposition, and the political economy analyses. The social support viewpoint contends that informal social networks are the fundamental determinants of both objective and subjective well-being. The inequality proposition intimates that the increasing economic disparities across the social ladder have adversely impacted people's sense of belonging and economic participation thereby triggering high anxiety conditions. The political economy perspective on the other hand argues that the main contributing factor to negative health outcomes is social and political exclusion exemplified in material deprivation. These perspectives viewed in isolation of each other reveals lack of scalar analyses for understanding the complexities in social capital and how that impacts health. Consequently, Szreter and Woolcock (2004) distinguishes between three forms of social capital, which are central to scale to reveal how other social elements impact health and human welfare. Bonding social capital refers to relationships of trust and cooperation between people who share social identity. Whereas bridging social capital implies a relationship between people who are not alike and may not share social identity and are based on mutual respect, linking social capital, the main intellectual addition by Szreter and Woolcock (2004) comprises the vertical social network and relationships between individuals and social institutions and political organizations. Poortinga (2012, p. 288) sums up this conceptualization brilliantly: "the different forms of social capital are conceptualized as being made up of 'ingroup' social cohesion and civic participation (bonding social capital), 'outgroup' social cohesion and heterogeneous relationships (bridging social capital), and political participation, efficacy, and trust (linking social capital)." In an attempt to explore the interconnection among social capital indicators and how such indicators are interconnected at the neighborhood scale, Poortinga (2012) developed aspects and indicators of social capital for each of the typologies described. An aspect of bonding social capital intimated is bonding social cohesion–neighborhood cohesion, trust, belonging, and civic participation. Indicators for measuring this aspect include the extent to which a neighborhood pull together to improve the community in question, perceptions on how many people could be trusted in the community, a strong sense of belonging, as well as involvement with groups and organizations. Social cohesion, mutual respect, heterogeneity of relationships are the aspects identified for bridging social capital. Indicators for bridging social cohesion include the extent to which people from diverse social and cultural backgrounds get on well together and respect ethnic differences, percentage of friends in similar socioeconomic status (income groups), as well as percentage of friends in of the same ethnicity. Four main aspects of linking social capital was proposed: political participation, political activism, political efficacy, and political trust. The proposed indicators comprise attendance of public meetings or rally or active participation in petition signing, perception of one's influence in local governance and decision-making, trust for local council, police, and elected officials. Empirical results from Poortinga's hypotheses showed that bonding social cohesion, civic participation, bridging social cohesion, heterogeneity of relations, and political efficacy were essential for health at the neighborhood scale.

The strength of social capital across scale impacts lifestyle and health outcomes of migrants. Strong social capital in the form of strong bonding and healthier lifestyles that are enshrined in culture at the destination impact health positively. This is because strong bonds act as protective structure for the newly arrived migrants to perpetuate the epidemiological paradox (Hosper et al. 2007); a prime example of the coupled social capital–health status relationship (Giordano et al. 2012; Moore et al. 2011). Major conduits by which social capital, when it is a protective factor, affects health of migrants include positive pressure on individuals for good behavior, access to health information and guidance, as well as emotional support (McPherson et al. 2006). In a study of social capital, lifestyle, and health outcome among migrants in Ghana, Tutu et al. (2017) found that receipt information about how to live successfully in an informal settlement was significantly associated with self-reported health status. Specifically, they found that migrants who receive information (including advice and guidance) from the host population (bridging social capital) regarding the way to live successfully in James Town, Accra, were more likely to rate their health better.

The interconnection between migration, social capital, and health is ubiquitous. While different forms of social capital have diverse impact on health outcome, the scale of analyses and the place of the study ought to guide the selection of variables for measuring the impact of social capital on health. That calls for the identification of specific social relations and the myriad pathways, including lifestyles, by which individuals and communities get engrossed in activities and decisions that ultimately influence health status and outcomes. More of this dynamic is discussed in Chap. 5.

References

Boyd M (1989) Family and personal networks in international migration: recent developments and new agendas. Int Migr Rev 23(3):638–670

Castles S, De Haas H, Miller MJ (2013) The age of migration: international population movements in the modern world. Macmillan International Higher Education, London

Chattopadhyay A, White MJ, Debpuur C (2006) Migrant fertility in Ghana: selection versus adaptation and disruption as causal mechanisms. Popul Stud 60(2):189–203

Fobil JN, Levers C, Lakes T, Loag W, Kraemer A, May J (2012) Mapping urban malaria and diarrhea mortality in Accra, Ghana: evidence of vulnerabilities and implications for urban health policy. J Urban Health 89(6):977–991

Fone DL, Dunstan FD, Lloyd K, Williams G, Watkins J, Palmer SR (2007) Does social cohesion modify the association between area income deprivation and mental health? A multilevel analysis. Int J Epidemiol 36:338–345

Giordano GN, Björk J, Lindström M (2012) Social capital and self-rated health—a study of temporal (causal) relationships. Soc Sci Med 75(2):340–348
Helliwell JF (2003) Well-being and social capital: does suicide pose a puzzle? Presented at the conference on well-being and social capital, Harvard, 7–9 Nov 2003. Available from http://www.wcfia.harvard.edu/conferences/socialcapital/happiness%20Readings/helliwell_2003.pdfS
Hollifield JF (2004) The emerging migration state 1. Int Migr Rev 38(3):885–912
Hosper K, Nierkens V, Nicolaou M, Stronks K (2007) Behavioural risk factors in two generations of non-Western migrants: do trends converge towards the host population? Eur J Epidemiol 22:163–172
Jankowska MM, Stoler J, Ofiesh C, Rain D, Weeks JR (2015) Agency, access, and Anopheles: neighborhood health perceptions and the implications for community health interventions in Accra, Ghana. Glob Health Action 8(26492):1–13
Kawachi I, Kennedy BP, Lochner K, Prothrow-Stith D (1997) Social capital, social inequality, and mortality. Am J Public Health 87:1491–1498
Kennedy BP, Kawachi I, Prothrow-Stith D, Lochner K, Gupta V (1998) Social capital, income inequality, and firearm violent crime. Soc Sci Med 47:7–17
Lu Y (2008) Test of the 'healthy migrant hypothesis': a longitudinal analysis of health selectivity of internal migration in Indonesia. Soc Sci Med 67(8):1331–1339
Lu Y (2010) Rural-urban migration and health: evidence from longitudinal data in Indonesia. Soc Sci Med 70(3):412–419
McKenzie K, Whitley R, Weich S (2002) Social capital and mental health. Br J Psychiatry 181:280–283
McPherson M, Smith-Lovin L, Brashears ME (2006) Social isolation in America: changes in core discussion networks over two decades. Am Sociol Rev 71(3):353–375
Moore S, Bockenholt U, Daniel M, Frohlich K, Kestens Y, Richard L (2011) Social capital and core network ties: a validation study of individual-level social capital measures and their association with extra-and intra-neighborhood ties, and self-rated health. Health Place 17(2):536–544
Palloni A, Morenoff J (2001) Interpreting the paradoxical in the Hispanic paradox: demographic and epidemiologic approaches. Ann N Y Acad Sci 954(1):140–174
Poortinga W (2006) Social capital: an individual or collective resource for health? Soc Sci Med 62:292–302
Poortinga W (2012) Community resilience and health: the role of bonding, bridging, and linking aspects of social capital. Health Place 18(2):286–295
Szreter S, Woolcock M (2004) Health by association? Social capital, social theory and the political economy of public health. Int J Epidemiol 33:650–667
Torun B, Schroeder D, Rodriguez M, Martorell R, Stein AD, Grajeda R et al (2002) Rural-to-urban migration and cardiovascular disease risk factors in young Guatemalan adults. Int J Epidemiol 31(1):218–226
Tutu RA, Boateng JK, Busingye JD, Ameyaw E (2017) Asymmetry in an uneven place: migrants' lifestyles, social capital, and self-rated health status in James Town, Accra. GeoJournal 82(5):907–921
Tutu RA (2013) Self-rated resilience among young migrants in old Fadama, Accra, Ghana. GeoJournal 78(4):709–725
Uchino BN, Cacioppo JT, Kiecolt-Glaser JK (1996) The relationship between social support and physiological processes: a review with emphasis on underlying mechanisms and implications for health. Psychol Bull 119:488–531
United Nations, Department of Economic and Social Affairs, Population Division (2015) Trends in international migration, 2015. Available from https://www.un.org/en/development/desa/population/migration/publications/migrationreport/docs/MigrationReport2015_Highlights.pdf. Accessed 29 Sept 2017

Chapter 3
Disease Patterns: Communicable and Non-communicable

Abstract The patterns of infectious and chronic diseases are discussed in this chapter. Drawing on the concept of the double burden of disease, a deliberation on the trends of communicable and noncommunicable diseases in the developing world is undertaken. A detailed portrait of the prevalence of these diseases in Ghana and Uganda is presented using secondary data sources. To ascertain whether the patterns observed at the national level is synonymous with the patterns in the slums, data are collected from the immediate clinics in the study communities. The data from the health facilities in all the four informal settlements illustrate the double burden of disease and echo the predominant diseases plaguing the two countries. The top ranked diseases responsible for years of life lost and living with disability in these countries including malaria, respiratory infections, diarrheal diseases, sexually transmitted diseases and other urinary tract infections, as well as hypertension. And these are experienced by residents in the poor communities. Body health mapping activity revealed that in addition to the disease mentioned, occupation-related illnesses contribute heavily to morbidity among the urban poor.

Keywords Communicable diseases · Noncommunicable diseases · Infectious diseases · Chronic diseases · Slum diseases · Uganda diseases · Ghana diseases · Health mapping · Disease burden · Double burden

Introduction

This chapter discusses the disease patterns observed and reported in the communities 4 months prior to the final survey on health literacy. Data were accessed from the medical facilities within the immediate neighborhood to understand the most common diseases. Tallies were made for the reported cases from the medical books given to us by the clinics as they were recorded and categorized based on objective diagnosis. The communicable diseases were new cases recorded, while the non-communicable diseases were both new and existing cases seen at the health facility during the month. These data informed our choices of communicable and noncommunicable diseases for migrants' health literacy to be discussed in Chaps. 5 and 6.

R. A. Tutu, J. D. Busingye, *Migration, Social Capital, and Health*,
Global Perspectives on Health Geography,
https://doi.org/10.1007/978-3-030-24693-8_3

Double Burden of Disease: Trends in the Developing World

The trends and patterns of diseases and their attendant mortality in the low-income countries illustrate the concept of double burden of disease (Agyei-Mensah and Aikins 2010). The concept explains that communicable and noncommunicable diseases coexist, and it is spatial in nature. Fundamentally, in the developing world, the spatial patterns of morbidity and mortality are a function of the co-occurrence of communicable and noncommunicable diseases and spatially differentiated. While the poor are more likely to experience high burden of communicable diseases and higher co-occurrence of communicable and noncommunicable diseases (Lopez et al. 2006), wealthier communities are more probable to have higher occurrence of chronic diseases (Oteng-Ababio 2014; Agyei-Mensah and Aikins 2010). Sub-Saharan Africa, especially, is experiencing an increase in noncommunicable disease epidemic including cardiovascular disease, cancer, and metabolic diseases like obesity and diabetes (WHO 2014; Dalal et al. 2011; Ezzati et al. 2005). Meanwhile, over 80% of all global deaths arising from noncommunicable diseases happen in low-to-middle income countries (Abegunde et al. 2007), and age-specific hypertension prevalence rates are highest in sub-Saharan Africa (46%) (WHO 2014); a testament to the assertion that poor communities bear the bigger weight of the double burden of disease (Kushitor and Boatemaa 2018). To add insults to injury, in the developing world, a lot of people living with chronic ailments such as hypertension and diabetes are not on medication because of inadequate access to facilities and the cost associated with treatment (Kolling et al. 2010). While longevity contributes immensely to the increasing at-risk population of noncommunicable diseases in sub-Saharan Africa (Fuster and Kelly 2010), other reasons are related to the rate of urbanization and changing lifestyles related to socioeconomic development (Vorster 2002; Steyn et al. 2005). The lifestyle factors include diet, smoking, physical inactivity, and alcohol use (Dalal et al. 2011). Although majority of the deaths in sub-Saharan Africa in 2004 was due to communicable diseases (malaria, HIV/AIDS and so on), 25% of all mortality in the region in the same year was due to noncommunicable diseases. The World Health Organization (WHO) projects that by 2030, noncommunicable diseases will be responsible for 46% of all deaths in the region (WHO 2005). Chief among these diseases causing death are cardiovascular diseases, chronic respiratory diseases, nutritional/endocrine disorders, and diabetes.

Disease Trends in Ghana and Uganda

Regarding trends, we begin with a brief discussion on the patterns of disease in Ghana and Uganda at the national scale. Prominence is given to disease ranking by categories, causes of mortality, and risk factors. Mostly, the causes of premature death in Ghana have been classified into three namely (1) communicable, maternal, neonatal, and nutritional disease; (2) noncommunicable diseases; and (3) injuries

(Institute for Health Metrics and Evaluation (IHME) 2018a, b). Between 2007 and 2017, regarding communicable diseases, seven main causes of premature death were identified. These are malaria, neonatal disorders, HIV/AIDS, lower respiratory infections, diarrheal diseases, tuberculosis, and meningitis. With respect to non-communicable diseases, three most prominent cause of premature death are stroke, congenital defects, and ischemic heart diseases. Road injuries was the leading cause of premature death in the injury category.

Communicable diseases were the leading cause of premature death in Ghana within the same period. In 2007, malaria was the leading cause of death followed by neonatal disorders, HIV/AIDS, lower respiratory infections, diarrheal diseases, tuberculosis, and meningitis in that order. However, in 2017, malaria was ranked second leading cause of death with a percentage of negative 42.2%, and neonatal disorders became the number one leading cause of death. Tuberculosis and meningitis ranked ninth and eleventh leading cause of death, respectively. Malaria is still a major cause of death due to, among other reasons, the relatively lower proportion of households using insecticide-treated mosquito nets (ITNs). Sixty-eight percent of households in Ghana own an ITN. In urban areas, 78% of households own ITN compared to the 60% of households in rural areas (GSS et al. 2014). While HIV prevalence rate decreased in 2013, the prevalence rate among women in their reproductive ages was higher than the national average (GSS et al. 2014).

Concerning noncommunicable diseases, it was the second leading cause of death in the major categories. From the Ghana Demographic and Health Survey 2014, noncommunicable diseases and lifestyle behaviors such as high blood pressure (hypertension) and smoking of cigarettes are the most common adult health and lifestyle risk factors in Ghana (GSS et al. 2014). Stroke remained the leading cause of mortality among the noncommunicable diseases. This was followed by ischemic heart disease with a positive percentage change of 14.6% and congenital defects, respectively. This situation is further compounded by the unsatisfactory provision of noncommunicable disease diagnosis and treatment (Kushitor and Boatemaa 2018). Kushitor and Boatemaa (2018) found that, in Ghana, support for hypertension and diabetes care lacked significant attention by health sector and attention is overly focused on communicable disease—albeit very crucial as well.

These diseases are associated with most causes of years lived with disabilities in the country. Communicable, maternal, neonatal, and nutritional diseases are linked with health problems that cause disability. These health problems are dietary iron deficiency, vitamin deficiency, and neonatal disorders, which were ranked (in terms of severity) first, fourth, and eighth, respectively, in 2007. These problems were still among the top 10 causes of years lived with disability in 2017, although vitamin A deficiency had a negative percentage change of 15.4% and neonatal disorders had a positive change of 52.6% thereby moving up to rank sixth.

Health problems related to noncommunicable diseases responsible for the years of life lost and living with disability in Ghana are headache disorders, lower back pain, depressive disorders, diabetes, age-related hearing loss, COPD, blindness and vision impairment, and anxiety disorders. From 2007 to 2017, the problems with an increasing contribution to years of life lost were depression disorders, diabetes, and anxiety disorders.

IHME (2018a) identified three broad primary risk factors that drive most death and disability. These are metabolic risks, environmental and occupational risks, and behavioral risks. Out of 10 ranked risk factors of the broad risk factors, behavioral risk factors dominated. These were malnutrition, unsafe sex, dietary risks, alcohol use, and tobacco use. The environmental risk factors include poor water, sanitation, and hygiene conditions, as well as air pollution. Metabolic risk factors constituted high blood pressure, high fasting plasma glucose, and high body mass index. In 2017, the behavioral risk factors with greater increasing contribution to adverse health outcomes were alcohol use and dietary risks. High blood pressure seems to have a higher contribution to disability adjusted years with a percentage change of 17% from 2007 to 2017.

Like Ghana, causes of premature death in Uganda are communicable diseases, noncommunicable diseases, and injuries. Nine of the most common communicable diseases responsible for premature deaths in Uganda are malaria, HIV/AIDS, neonatal disorders, lower respiratory infections, diarrheal diseases, tuberculosis, meningitis, protein-energy malnutrition, and other sexually transmitted diseases (IHME 2018b). While congenital defect is the major noncommunicable disease causing premature death, road injury is the predominant injury causing mortality.

Between 2007 and 2017, there has been changes in the contribution of each of the diseases to mortality. While HIV/AIDS was the highest contributor to premature death in 2007 in Uganda, it became the second highest contributor in 2017—a negative percentage change of 64%. Neonatal disorders were ranked one in 2017 from third in 2007 and malaria ranked third in 2017 from second in 2007. The reduction in malaria's contribution to mortality in Uganda is attributed to, among other things, the increased household ownership of insecticide-treated nets. In 2016, 80% of households in Uganda owned one regular mosquito net (not treated with insecticide), and 78% owned an ITN (UBOS and ICF 2018). The 2016 Uganda Demographic and Health Survey results intimate that the percentage of households with ownership of an ITN increased from 16% in 2006 to 78% in 2016. Congenital defects maintained the fifth rank between the same period while road-side injuries ranked 10 in 2017 from 11th in 2007 with a positive percentage change of 2.3%.

Communicable and nutrition-related diseases that are among the top 10 causes of years of life lost are dietary iron deficiency, HIV/AIDS, vitamin A deficiency, schistosomiasis, malaria, and neonatal disorders. Anemia leads to the increase of maternal mortality and poor birth outcomes in pregnant women. Data from 2016 Uganda Demographic and Health Survey show that 32% of women in their reproductive ages had some degree of anemia—25% were mildly anemic, 6% were moderately anemic, and 1% severely anemic (UBOS and ICF 2018). Noncommunicable diseases on the other hand include depressive disorders, low back pain, headache disorders, and anxiety disorders. In 2017, three of the top five causes of years of life lost were noncommunicable diseases—depressive disorders, low back pain, and headache disorders. The causes of most deaths and disability combined in 2017 in

Uganda were mostly communicable diseases. The first five highest ranks were neonatal disorders, HIV/AIDS, malaria, and lower respiratory infections. There were 1.4 million of adults and children living with HIV in Uganda as of 2016 (UBOS and ICF 2018). According to the United Nations AIDS (UNAIDS 2017), Uganda has approximately 52,000 new HIV-infected cases and 28,000 HIV-related deaths annually.

Like Ghana, metabolic risks, environmental risks, and behavioral risks were the risk factors that drive most deaths and disability in Uganda. Behavioral risk factors dominated, and their rankings from highest to lowest contributor in 2017 were malnutrition, unsafe sex, alcohol use, dietary risk, and tobacco use (IHME 2018b). Regarding environmental risk factors, water, sanitation, and hygiene were the highest contributor followed by air pollution and occupational risks. The metabolic risk factors were high blood pressure and high fasting plasma glucose.

Disease Patterns: James Town and Old Fadama

In James Town, the community clinic serving the Ngleshie neighborhood is the Seaview Clinic. Although the Usher Fort Polyclinic is the major hospital in the area, this clinic seems to be the first point of call for many residents when they are sick. Not surprising, malaria cases were the highest in each month in both communities in Ghana with over 800 new cases in James Town and almost 400 cases in Old Fadama (Tables 3.1 and 3.2). Within the same period, urinary tract infection cases reported were 59 in James Town and over 200 in Old Fadama. Upper respiratory

Table 3.1 Number of recorded cases of diseases in Seaview Clinic, James Town

	Months				
Diseases	June	July	August	September	Total
Malaria	169	201	256	208	834
Hypertension	13	20	17	13	63
Diabetes	14	16	21	20	71
STI/UTI	14	15	21	9	59
Anemia	13	18	20	11	62

Table 3.2 Number of recorded cases of diseases in Star Aid Clinic, Old Fadama

	Months				
Diseases	June	July	August	September	Total
Malaria	76	88	99	130	393
Hypertension	30	15	17	21	83
URTI	59	67	30	19	175
UTI	51	35	63	65	214
Enteric fever	69	65	70	72	276

tract infection (URTI) cases were 175 in Old Fadama, and the health personnel in the clinic we interacted with suggested that URTI is the major public health concern in the community due to the physical environment. Hypertension and diabetes were the most common noncommunicable diseases reported in James Town, and personal communication with the medical staff revealed their concern for the seemingly gross lack of disease knowledge on hypertension. In Old Fadama, hypertension was the most common noncommunicable disease.

Disease Patterns: Namuwongo and Kinawataka

In Namuwongo, Joy drug shop and Ribbe clinic were the two health units that served as the health facilities that are visited by the community members. The immediate clinic serving Kinawataka residents is Mulongo clinic. The two units are all privately owned and operated by low level health qualified health workers; one by a retired nurse and the other by a midwife. There is a City Council Health Unit in Namuwongo; however, residents indicated that it was far from them and the response to an emergency situation was not as fast as it was with the small health units. Diarrhea cases were reported as the highest in each month in both communities with 1165 total new cases in Namuwongo and 1164 total new cases in Kinawataka for the 4 months tallied (Tables 3.3 and 3.4). In the same months, the clinic recorded 718 new malaria cases, while 199 hypertension cases (both new and old) were recorded in Kinawataka. Noteworthy is that during the months of August and September, the slums record high amounts of rainfall, which results into

Table 3.3 Number of recorded cases of diseases in Joy drug shop, Namuwongo

	Months				
Disease	June	July	August	September	Total
Diarrhea	202	256	306	401	1165
Malaria	105	109	201	303	718
Hypertension	50	52	56	58	216
STDs	15	27	41	56	139
Ring worm	14	23	25	42	104

Table 3.4 Number of recorded cases of diseases in Mulongo Clinic, Kinawataka

	Months				
Disease	June	July	August	September	Total
Diarrhoea	205	260	301	398	1164
Hypertension	42	48	54	55	199
Malaria	35	39	45	48	167
STDs	22	25	29	31	107
Allergies	10	15	18	22	65

flooding. Health workers reported that with flooding, many cases of diarrhea are reported by members of the community.

Brief Deliberation on Disease Patterns

The data from the health facilities in all the four informal settlements illustrate the double burden of disease being experienced by the poor in the developing world. The diseases do not only mirror the coupled communicable and noncommunicable disease burden but also reflect the predominant diseases plaguing the two countries. The top ranked diseases responsible for years of life lost and living with disability in these countries including malaria, respiratory infections, diarrheal diseases, sexually transmitted diseases and other urinary tract infections, as well as hypertension are experienced by residents in these poor communities.

More so, the uniqueness of place and disease distribution is exemplified as well. Apart from the variation in the number of reported cases, some diseases did not feature in the top five most reported cases. For example, for the informal settlements in Ghana, while malaria and urinary tract infections were common to both slums with varying degrees of reported cases, enteric fevers and respiratory tract infections were unique to Old Fadama in terms of featuring in the top five most reported cases. The high rate of enteric fever in Old Fadama, according to the health care staff in the Old Fadama clinic, is due to typhoid fever. As discussed previously regarding the environmental state of the slum, including the poor hygienic surroundings, the presence of enteric fever (typhoid) is very plausible. We proceeded to examine how these patterns are reflected among migrants in this community using participatory body health mapping activities.

Body Health Mapping

Body health mapping activities with migrant community members to understand self-reported diseases were undertaken. Participants drew a life-sized outline of the human body on a large piece of paper. One-by-one they were asked to indicate the various body parts where they were feeling pain/hurt or had health problems by using color-coded self-adhesive dots. Sex was differentiated by color. The activities were anonymous. That is, the life-sized human body was hidden behind a tree/wall/screen before participants indicated the health problems. The emerging patterns were subsequently discussed. Participants were asked to report and point out the symptoms and diseases based on the final product (Fig. 3.1) without necessarily discussing their particular or individual diseases. Perceived reasons and causes of the diseases were elicited as well. During the discussions, attention was paid to patterns of communicable and noncommunicable diseases, occupation-related diseases, spatial patterns, and clustering patterns.

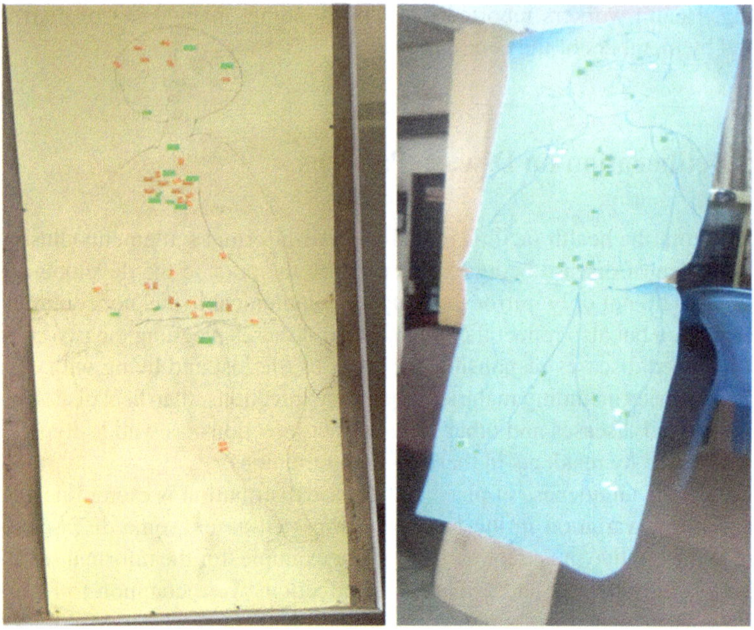

Fig. 3.1 Body health mapping (Old Fadama—left; James Town—right)

Old Fadama and James Town, Ghana

In Old Fadama, three main clusters were observed. The color-coded self-adhesive dots were concentrated around the head, chest, and waist/abdomen. Regarding the cluster around the head, participants mentioned frequent fevers, specifically malaria and typhoid, to be associated with headaches they frequently feel. While malaria and other fevers were unanimous, high blood pressure was mentioned by a couple of participants. The residential location of the individual in the slum was not perceived as having influence on who get malaria or not. On the clusters around the chest, the discussions revealed two main patterns. Firstly, the experience of diseases that are related to their immediate environment, and secondly, occupation-related health issues. Participants mentioned that they usually feel pain at the chest when they experience constant cold, coughing, and asthma. For participants who are involved in manual labor, complaints about chest pains were also linked to the carrying of heavy load on their heads in the hot sun all the time. See Figs. 3.2 and 3.3 for picture of head porterage and metal scrap dealing—quintessential of manual labor. With respect to abdominal pains, participants mentioned drinking too much alcohol, eating undercooked foods, experiencing menstrual cramps, and suffering from ulcer as responsible for stomach aches and abdominal pains. Related to the waist area, issues like hemorrhoids (pile), spinal cord problem, and the discharge of blood and other materials from the lining of the uterus were mentioned.

Fig. 3.2 Head porters in Old Fadama (authors' fieldwork)

Fig. 3.3 Metal scrap dealership in Old Fadama

While a similar clustering pattern is observed on the James Town body health map except for abdomen, the discussions revealed some differences. Clustering around the head was attributed to eye problems including glaucoma resulting from watching television in unsafe proximity, headaches due to stress and onset of high blood pressure, as well as light rays from mobile phones at night. The concentration around the chest was mainly associated with the hazards of their occupations and lifestyles. The reasons mentioned include carrying of heavy items, the lifting of heavy items, hypertension, and lifestyle behavior such as smoking. Probing for more information on hypertension, participants mentioned that, apart from stress, excessive alcohol consumption and smoking as reasons for developing the disease. Thus, unlike Old Fadama, migrants reported less upper respiratory infections and none of the reasons assigned for the clustering on the head and chest areas related to URI, a further illustration of place characteristics impacting perceptions of disease. For waist pains, participants mentioned sitting for long hours, walking long distance, lifting heavy items, and standing for too long due to the nature of economic activities they are engaged in such as trading or food vending.

Namuwongo and Kinawataka, Uganda

In Kinawataka, there were three main clusters identified (Fig. 3.4). The color-coded self-adhesive dots were concentrated around the genital organs, the abdomen, and the head. The clusters around the abdomen, according to the migrants, represented diseases such as diarrhea, constipation, dysentery, and excessive gas. These diseases were said to be related to where one lived in the community and to the weather patterns, and incidences escalate during the rainy seasons. The cluster around the abdomen was representative of STDs/STIs, smelly discharge, itching, thrush, swelling of the penis, and low sexual energy. Migrants often referred to all diseases or infections around the genitalia as syphilis or gonorrhoea and candidiasis. This was the most interesting discussion with migrants accusing each other of engaging in

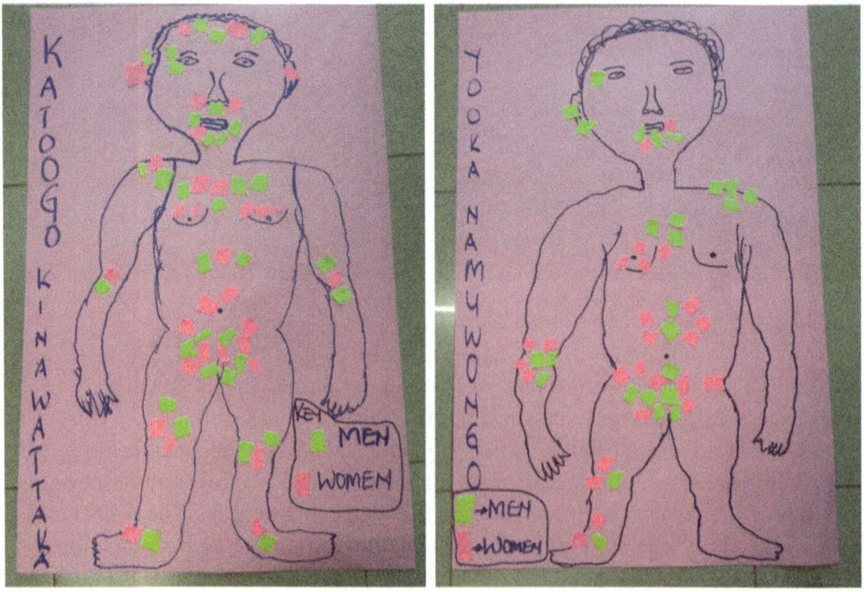

Fig. 3.4 Body health mapping (Kinawataka—left; Namuwongo—right)

unprotected sexual encounters, hence spreading diseases. There was a general concern that these days when they visit health units, they are told that they have an infection, without clearly stating what the disease is. Our personal conversations with health workers in the clinics supported the assertion of the migrants. The health workers indicated that this happens because they sometimes are not able to explain the terms in the local languages. The cluster around the head, participants reported, was mainly due to headaches caused by hypertension and stress because of too many thoughts related to daily life experiences.

While there were no significant differences in the body mapping in the two places, in Namuwongo, the concentration of the dots was around the waist/stomach, abdomen, and genitalia. The dots around the waist, according to the migrants, were a depiction of the infections and diseases that included diarrhea, dysentery, and typhoid. Here, instead of focusing on the anal part, to represent such diseases associated with watery stool and frequent visits to the toilet, participants chose to focus on the waist or stomach area. Participants indicated that these diseases are associated with consumption of dirty water, eating cold contaminated food, and eating with unwashed hands. The dots around the abdomen represented the lower abdominal pain associated with painful menstrual periods, cancers, and other associated illnesses. Participants also reported that the dots around the genitalia were representative of STIs, STDs, and UTIs. They indicated that due to the poor hygiene of the latrines they use, they often contract UTIs and experience a lot of pain while passing urine. Others reported that they get severe itching around the genitalia and get swellings and wounds. In addition, they also experience gonorrhoea and the urine passage gets blocked. See Fig. 3.5 for a picture of migrant participants in Namuwongo during a body health mapping activity.

Fig. 3.5 Participants of body health mapping activity, Namuwongo (authors' fieldwork)

Brief Deliberation on Body Health Mapping

The usual disease suspects, especially malaria, seem to be a leading source of morbidity among migrants in the communities in Ghana. Apart from the diseases revealed through the data from the health centers, occupation-related stressors appear to weigh heavily on the migrants and contributing to varied forms of morbidity especially in Old Fadama. In the communities in Uganda, diarrheal diseases among migrants reflected the dominant diseases reported in the health centers. Clearly, infectious diseases (malaria, respiratory infections, sexually transmitted diseases, and infection) are still the major reason for morbidity among the poor in the urban periphery, but as evidently indicated, the double burden of disease is very much in play with reference made to hypertension across all the slums.

References

Abegunde DO, Mathers CD, Adam T, Ortegon M, Strong K (2007) The burden and costs of chronic diseases in low-income and middle-income countries. Lancet 370:1929–1938

Agyei-Mensah S, Aikins ADG (2010) Epidemiological transition and the double burden of disease in Accra, Ghana. J Urban Health 87(5):879–897

Dalal S, Beunza JJ, Volmink J, Adebamowo C, Bajunirwe F, Njelekela M, Mozaffarian D, Fawzi W, Willett W, Adami HO, Holmes MD (2011) Non-communicable diseases in sub-Saharan Africa: what we know now. Int J Epidemiol 40(4):885–901

Ezzati M, Vander Hoorn S, Lawes CM et al (2005) Rethinking the ‘diseases of affluence’ paradigm: global patterns of nutritional risks in relation to economic development. PLoS Med 2:e133

Fuster V, Kelly BB (eds) (2010) Promoting cardiovascular health in the developing world: a critical challenge to achieve global health. The National Academies Press, Washington, DC

Ghana Statistical Service (GSS), Ghana Health Service (GHS), and ICF International (2015) Ghana demographic and health survey (2014) GSS, GHS, and ICF International, Rockville

Institute of Health Metrics (2018a) Country profile Ghana [Internet]. Available from http://www.healthdata.org/ghana. Accessed 6 Jun 2018

Institute of Health Metrics (2018b) Country profile Uganda [Internet]. Available from http://www.healthdata.org/uganda. Accessed 6 Jun 2018

Joint United Nations Programme on HIV/AIDS (UNAIDS) (2017) UNAIDS data 2017. UNAIDS, Geneva

Kolling M, Winkley K, von Deden M (2010) “For someone who’s rich, it’s not a problem.” Insights from Tanzania on diabetes health-seeking and medical pluralism among Dar es Salaam’s urban poor. Glob Health 6:1–9

Kushitor MK, Boatemaa S (2018) The double burden of disease and the challenge of health access: evidence from access, bottlenecks, cost and equity facility survey in Ghana. PLoS One 13(3):e0194677. https://doi.org/10.1371/journal.pone.0194677

Lopez AD, Mathers CD, Ezzati M, Jamison DT, Murray CJ (2006) Global and regional burden of disease and risk factors, 2001: systematic analysis of population health data. Lancet 367:1747–1757

Mathers CD, Loncar D (2006) Projections of global mortality and burden of disease from 2002 to 2030. PLoS Med 3:e442

Oteng-Ababio, M. (2014). Rethinking waste as a resource: insights from a low-income community in Accra, Ghana. City, Territory and Architecture, 1(1), 10

Steyn K, Sliwa K, Hawken S et al (2005) Risk factors associated with myocardial infarction in Africa: the INTERHEART Africa study. Circulation 112:3554–3561

Uganda Bureau of Statistics (UBOS) and ICF (2018) Uganda demographic and health survey 2016. UBOS and ICF, Kampala

Vorster HH (2002) The emergence of cardiovascular disease during urbanization of Africans. Public Health Nutr 5:239–243

WHO (2014) Global status report on noncommunicable diseases 2014. Geneva. ISBN 9789241564854

World Health Organization (2005) Global burden of disease. Projections of mortality and burden of disease, 2002–2030. http://www.who.int/topics/global_burden_of_disease/en/

Chapter 4
Environmental Risks to Health

Abstract This chapter focuses on environmental risk to health in the study sites. Environmental exposures and contaminants are assessed through testing of chemical and biological contaminants using test strips, undertaking participatory hazard mapping activities on perceived dangerous and safe places for migrants' health, and survey on the perception of the effects of the environment on food safety and its correlation with receipt of foodborne diseases information. Results from the chemical and biological contaminant tests show that many households sampled in the slums in both countries have high levels of heavy metals and bacteria pollutants in their nondrinking water that may be dangerous for health. The participatory hazard mapping activities showed migrants' perceptions of their places of residence and environmental risks to health. Place characteristics and its relationship with perceptions of environmental risk to health were apparent. For instance, the presence of the unkept surroundings of the abattoir in James Town determined the kinds of risks and associated diseases, whereas the soap factory in Kinawataka influenced perception of risks in that place of residence for migrants. However, we observe that migrants demonstrated limited knowledge of diseases that may be associated with the environmental hazards in their neighborhoods.

Keywords Health risks · Environmental risks · Chemical contaminants · Slum pollutants · Bacteria contaminants · Hazard mapping · Foodborne diseases · Risks perception · Urban health · Water contaminants

Introduction

In this chapter, we describe environmental risk to health in the study sites with emphasis on the types of environmental exposures and contaminants. These are achieved through (1) test of chemical and biological contaminants in the environment in all four slum communities in Ghana and Uganda; (2) undertaking

R. A. Tutu, J. D. Busingye, *Migration, Social Capital, and Health*,
Global Perspectives on Health Geography,
https://doi.org/10.1007/978-3-030-24693-8_4

participatory hazard mapping activities on perceived dangerous and safe places for migrants' health in all the four slums in both countries; and (3) activities on the perception of the effects of the environment on food safety and its correlation with receipt of foodborne diseases information, using survey results from a community in Ghana. The chapter shows the impact of place on the perceptions of environmental vulnerability.

A large proportion of the global burden of disease has been attributed to poor environmental conditions and therefore ensuring that environmental services are properly maintained underpins varying aspects of human health (WHO 2008, 2016). To improve global health will mean to reduce environmental risks to health (WHO 2016; Ezzati et al. 2005), especially in urban informal settlements (Brown et al. 2018). One hundred and one out of the 133 disease and disease groups enumerated on the Global Health Observatory are significantly environment related (WHO 2016). While environmental factors are the primary cause for a quarter of the global burden of disease, it is responsible for a third of the burden of disease in the developing world, especially in sub-Saharan Africa (Prüss-Üstün and Corvalán 2006). Twenty-three percent of global mortality are attributed to modifiable environmental features (WHO 2016). Consequently, actions can be taken to decrease environment-related disease burden. Among such actions are more cautious use of environmental contaminants in the farm, home, and industrial sites, improved hygiene and household water consumption measures, the promotion of the use of purer and safer fuels, as well as implementing sanitation protocols.

The pathways by which environmental conditions impact health and health outcomes are multiple and varied (Briggs 2008); consequently, a number of frameworks have been used to explain the linkages and to seek avenues for redress. These include environmental hazards, ecosystem approach, and vulnerability. Regarding environmental hazards, this is framed around environmental risk factors and their impact on health. That is, what kinds of diseases and how much of such diseases are because of specific environmental hazards such as polluted household drinking water and thereby identifying measurable goals when the hazard is addressed. It has been viewed from disease outcome perspective as well. That is, identifying the percentage of a disease that is the outcome of an environmental hazard or caused by synergistic effects of two or more environmental hazards. There is evidence of health ramification of hazardous waste exposure (Fazzo et al. 2017).

In a framework illustrating exposure to environmental risk factors, Ezzati et al. (2005) asserted that identifying proximal factors such as waste disposal infrastructure, access to clean water, and presence of pathogens in drinking water is important to understanding the underlying risks factors for specific diseases. Research has revealed that key environmental risk to health with measurable disease effects are unsafe water and sanitation, lead exposure, indoor air pollution, urban air pollution, and climate change. According to the World Health Organization (WHO), globally, almost 2 million people die yearly from unsafe water and sanitation due to a range of waterborne disease. Lead exposure results in cardiovascular diseases associated with hypertension and childhood mental retardation leading to 13 million disability-adjusted life years; indoor air pollution from solid fuel use and associated diseases was responsible for 1.5 million deaths in 2002. Owing to increased levels of fine particles in ambient air from industry and vehicles, there is an estimated increase in cardiopulmonary diseases and acute respiratory infections, which may be responsible for an estimated 800,000 every year. For the developing world context, studies have shown which particular agents and vectors of environmental risk factors are of relevance (Ezzati et al. 2005). For water and sanitation, fecal contamination levels, level of pathogens, as well as the concentration of dangerous chemicals in food and water, on the ground and hands. On indoor air pollution, pollutants concentration is a major indicator. These pollutants include carbon monoxide, fuel chemical composition—arsenic and fluorine, and respirable particles. Ambient concentration and chemical characteristics of pollutants in multiple urban spaces is the recommended indicator for urban air pollution. With malaria as one of the major public health issues, the agents and vector of relevance suggested including entomological inoculation rate and parasitemia above certain limits. Meanwhile, because low-income countries have more illegal dumping sites due to poor environmental management laws and policies, the presence of hazardous waste has been found in landfills unknowingly and has been found to be linked to diseases such as liver, bladder, breast and testis cancers, non-Hodgkin lymphoma, and asthma (Fazzo et al. 2017).

Additionally, there are diseases with significant environmental influence. These are diarrheal diseases, unintentional poisoning, lower respiratory infections, road traffic injuries, as well as vector-borne diseases. WHO estimates that 94% of the 1.8 million yearly deaths from various diarrheal disease is as a direct result of environmental conditions such as polluted drinking water and unsanitary conditions. Deficient management of chemicals in the agricultural sector as well as other occupational settings occasions 71% of all unintentional poisonings and results in about 350,000 deaths yearly, and most of these deaths occur in the developing world. Environmental factors such as poor residential settings, ecosystem degradation, and poorly designed irrigation systems have become safe havens for disease vectors. So, yearly, these result in over 500,000 deaths, mostly from malaria. While lower respiratory infections (e.g., pneumonia, bronchitis, and bronchiolitis accounted for 935,000 deaths per year in 2013) are acquired from exposure to indoor smokes from solid fuels, environmental factors like poor transport infrastructural design and land-use design make people susceptible to traffic accidents leading to over 467,000 deaths from injuries sustained due to road accidents.

Concerning ecosystem's approach, a nexus of relations is established between the environment and health that are critical to both individual and institutional agency. The contributions to health by ecosystems are vital in many respects. Ecosystem services ensure human sustenance through the provision of food, medicine, shelter, purification of the air, soil fertility, and regulation of disease vectors and pathogens. In other words, ecosystem is essential for the regulation of disease transmission with direct impact on global burden of disease. Therefore, the degradation of ecosystems is expected to impact health in a myriad of ways including increased malnutrition from soil degradation, rising diarrheal diseases due to contamination of water bodies, and increased vector-borne disease because of poor environmental management regimes.

Some vulnerability frameworks have focused on vulnerable populations and their environments of abode. Studies have shown that children are more susceptible to vector-borne disease like malaria as well as diarrheal diseases that are related to contaminated drinking water and unhygienic environments. Out of the 12 million environmental-related deaths among children globally, 4 million of such deaths are children aged 14 years and below (Prüss-Üstün and Corvalán 2006). Additionally, the poor are known to be at greater risk from the health-related impact of poor environmental management like pollution. The burden of lower respiratory infections is particularly high among the global poor due to over reliance on solid fuels for cooking and, hence, exposure to indoor smoke and pollutants. Poor subsistence farmers and industrial workers are more likely to misapply chemicals and work in environments polluted by industrial chemicals and pollutants thereby affecting their health adversely.

Migration into slums, as described previously, is associated with settling in peripheral neighborhoods of contamination. Urban informal settlements are home to about a million people with insufficient water and sanitation amenities. The inadequacy leads to access to contaminated water, poor sanitation, and hygiene, which is responsible for about 80% of disease and a quarter of all deaths in the developing world (WHO 2017; Brown et al. 2018). Globally, many migrants have willingly and unwillingly moved into and lived in places with contaminated lands including places with leaked industrial waste and severe environmental risk to health (Filcak 2010). Marginalized and poor migrants who flee desperately from conflicts, for example, may not have the wherewithal to scout for descent accommodation or neighborhoods to settle through conscious settlement decision and, therefore, will make do with whatever settlement is available. Consequently, such migrants end up settling in places where they are exposed to environmental risk to health such as contaminated water and air (Filcak 2010). The contamination in the informal settlements worsens the risk of exposure and exalts the inextricable connection between adverse environmental conditions and negative health outcomes (Brown et al. 2018). Migration has also been conceived as a mediator between environmental risk to health and health outcomes. It has been established that migration is connected to mood and anxiety disorder due to environmental factors. For example, migration into densely populated environments has been found to be linked with high risk of developing schizophrenia (Vilain et al. 2013). Also, concerns have been expressed about the potential of Chagas disease spreading to nonendemic countries because of human migration. This is possible because if human carriers of the parasite migrate and encounter potential vector species at the destination, then that may be the point of the presence of the disease in nonendemic places (WHO 2016). Migration is known to be responsible for the spread of vector-borne diseases. While leishmaniasis vectors in Central and South America are usually found in natural environments like forest, due to deforestation and migration, they have spread even into cities with increasing transmission to humans around homes (WHO 2016).

Given the established link between the environment and health, we proceeded to test for contaminants in the environment in our study sites. What are the safe spaces and dangerous places for health according to migrants? Are contaminants on their radar? If so, which are the ones that are commonly or rarely mentioned? How do migrants perceive their environment in relation to health, especially regarding food?

Test of Chemical and Biological Contaminants

To explore the presence of chemical and biological contaminants in the slums, household nondrinking water from at least 50 households in each of the communities was tested. This was a subsample of the 150 participants from each of the slums who participated in the study. The households were selected across the length and breadth of each slum. Chemical indicator strips were used to test the water.

To test for chemical contamination in household nondrinking water, indicator strips, which are colorimetric methods designed to detect the presence of different heavy metals in water samples, were used. Specifically, Sen Safe Water Metals Check kit was used (Fig. 4.1). It can detect the presence of heavy metals that may be of health concern such as Cadmium, Cobalt, Copper, Iron, Lead, Mercury, Nickel, and Zinc. A strip is dipped into a 20-ml water sample for 30 s with a constant back and forth motion; after 2 min, the color of the circle on the strip is visually compared

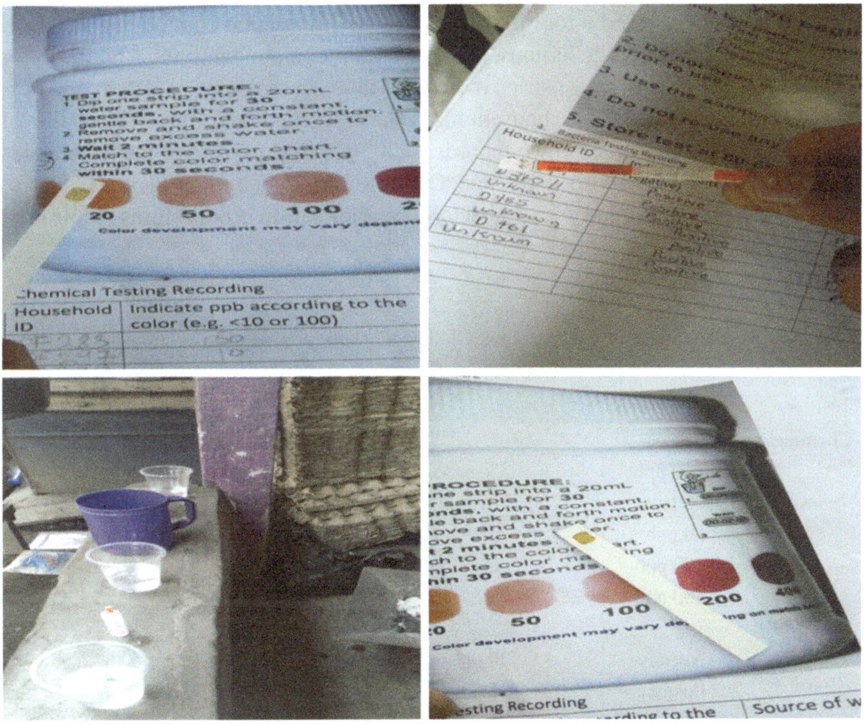

Fig. 4.1 Chemical and biological testing in Old Fadama and James Town

to the reference scale provided, ranging from <10 to 400 parts per billion (ppb) (μg/L). Ideally, it is expected that the indicator after the test should read <10 ppb.

Biological contamination in household nondrinking water was tested using *Watersafe* bacteria test strips. This kit contains a pipette, a vial, and a test strip. It can detect various types of bacteria such as *E. coli*, *Pseudomonas aeruginosa*, species of Shigella, Enterobacter, coliform, and noncoliform bacteria, which may be harmful to human health. A clean pipette is used to extract water sample into a vial. The vial is swirled and allowed to sit for 5 min after which the test strip is dipped into the vial. After 10 min, reddish lines appear on the test strip. Lines appearing only on the control line (C or 2) indicate a negative result while two lines appearing on both the test and control (T or 1 and C or 2) lines indicate a positive result. Information on the source of nondrinking water, storage device, and device used to fetch water sample for testing were elicited. Storage devices were categorized as direct from the pipe, drum/barrel covered, drum/barrel uncovered, bucket, jerry cans, and plastic bottles. Devices used to fetch the water sample include metal cup, rubber cup, plastic bowl, silver bowl, and cooking pan.

James Town and Old Fadama Results

The testing kits provided a tool for measuring environmental contaminants as well as to both whet the appetite of households for water contaminants and a platform for discussion. The chemical test strips indicated values in the order of <10 to 400 ppb. In Old Fadama, 34% of the households tested recorded >20 ppb, while 40% of households tested in James Town recorded >20 ppb. Five percent of the households tested in James Town recorded 400 ppb. On biological contaminants testing, 42% of households tested recorded positive bacteria contamination in both slums (Table 4.1).

Table 4.1 Results of contaminants testing, Old Fadama and James Town

		Old Fadama		James Town	
Test	Categories	*N*	%	*N*	%
Chemical					
	<10 ppb	33	66	35	59.3
	20 ppb	15	30	15	25.4
	50 ppb	2	4	4	6.8
	100 ppb	0	0	2	3.4
	400 ppb	0	0	3	5.1
Bacteria					
	Negative	29	58	34	57.6
	Positive	21	42	25	42.4
Total		50	100	59	100

Fig. 4.2 Water storage devices, sources, and environment in James Town

The results suggest that a substantial number of households may be using water that may contain chemical and bacteria levels that are dangerous to health, although for nondrinking purposes. See Fig. 4.2 for water storage devices, sources, and environment in James Town.

In both slums, there was no significant association between the source of water, storage device, device used in fetching water on one hand, and the results of both the chemical and biological tests on the other.

Namuwongo and Kinawataka Results

In Namuwongo, 81% of the houses tested recorded >50 ppb, while 70% of the households in Kinawataka recorded >50 ppb. Seventeen percent of the households tested in Namuwongo recorded high concentration of heavy metals in water at >100 ppb. On biological contaminant testing, 85% of households in Namuwongo and 72% in Kinawataka tested recorded positive bacterial contamination in both slums (Table 4.2). The results from both slums indicate that most of the households are using highly contaminated water by both heavy metals and bacteria. Worth noting is that most of the water tested was not necessarily for drinking; however, children were reported to sometimes consume the water.

In both Namuwongo and Kinawataka, there was no significant relationship between the source from which water was drawn, the water storage device, device used to draw water on one hand, and the level of biological and chemical contamination on the other.

Table 4.2 Results from contaminants testing, Namuwongo and Kinawataka

Test	Categories	Namuwongo		Kinawataka	
		N	%	*N*	%
Chemical					
	<10bbp	0	0	4	08
	20 ppb	1	2	11	22
	50 ppb	39	81	35	70
	100 ppb	8	17	1	02
	400 ppb	0	0	0	0
Bacteria					
	Negative	07	15	14	28
	Positive	41	85	36	72
Total		48	100	50	100

Brief Deliberation on Contaminants Testing

Regarding chemical contaminant testing, bearing in mind that the ideal water tested results should read <10 ppb, households in the slums in both countries have high levels of heavy metals that may be dangerous for health. The same can be said for the results of the bacteria tests. In our opinion, future research priority will be to isolate specific metals as well as examine the effects of exposure of the specific metal and bacteria and their corresponding health outcomes. For example, apart from the known associations that exist between specific metals and health outcomes like lead exposure leading to mental retardation in children, assessing precise contaminants in specific locations and the diseases associated with them will add to the existing knowledge of pollutants in the urban environment beyond ambient air and industrial contaminants.

Participants' direct involvement through water provision and observation of the tests enabled increased curiosity of the implications of the results. Each household participant was shown the results of each test and the reading explained, although the essence of the test was explained prior to the activity. While participants had been assured of the need not to be alarmed irrespective of the results, they were informed if the test results indicated levels that may be dangerous to human health. Although we had not examined the relationship between the source of water, storage devices, and item used to fetch the water on one hand and the test results on the other, households were admonished to ensure best water, sanitation, and hygiene practices always.

Participatory Hazard Mapping

This activity was undertaken for depiction of dangerous and safe places in community for migrants' health. Participants collectively brainstormed and drew a map (sketch) of the community on a large piece of paper. See Figs. 4.3 and 4.4 for

Fig. 4.3 Female participatory hazard mapping participants, Nawumongo

Fig. 4.4 Female participatory hazard mapping participants, James Town

participants' engagement during these activities in Nawumongo and James Town. Major landmarks in the community were indicated to use as points of reference during discussions. Migrant participants were asked to indicate dangerous places for health on the map using red/pink self-adhesive notes/dots. Drawings or sketches of what the "issues" are were made right next to the red/pink note/dots. Inquiries about the reason(s) for each red/pink sticky note/dot placed were made. Specifically, why is it a dangerous part of the community regarding migrants' health? Subsequently, participants were asked to indicate safe places on the map using green self-adhesive note/dots. Drawings or sketches of items/buildings/landmarks in the area that make the place safe were made. We inquired from migrants the reasons for each green sticky note/dot placed; that is, why is it a safe part of the community for migrants' health? Facilitators probed for linkages between dangerous places and different kinds of illnesses or diseases when participants did not mention any. Also, linkages between safe places and good health were queried. This activity was segmented by sex. Conceptually, health concerns differ between females and males. Also, spatial perception of dangerous and safe places for health may be substantially different among the sexes.

Participatory Hazard Mapping Results, James Town: Females

Migrants drew the map of the community with major landmarks such as the London market, Hotel, the public toilet, the roundabout, churches, police station, pharmacy shop, the Chiefs Palace, schools, and grocery shops (Fig. 4.5). Regarding places considered to be dangerous for health, female participants placed adhesive notes in places around a big gutter, the London market, the Hotel, and the public toilet. Migrants reported that the presence of the big drain near the London market makes one susceptible to diseases such as typhoid, malaria, cholera, candidiasis, and diarrhea. Participants mentioned that the drain represents other small drains in the

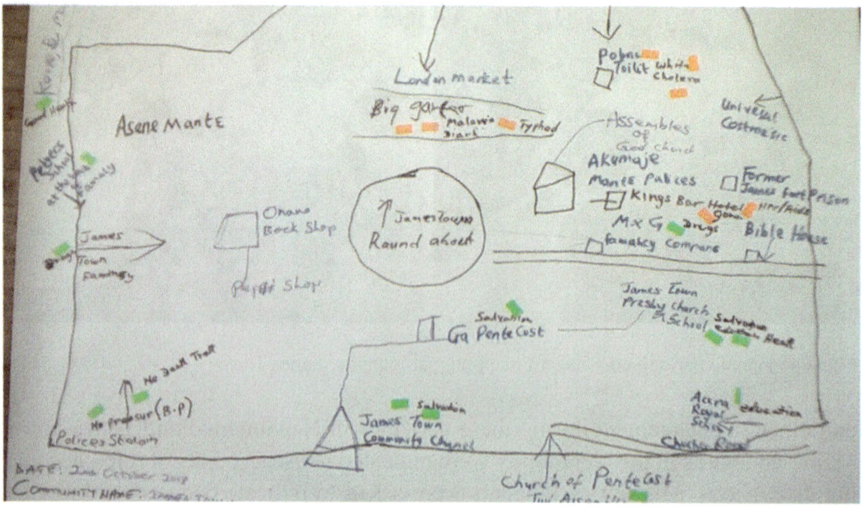

Fig. 4.5 Participatory hazard mapping, James Town (Females)

community that are unkept and hence the surrounding areas are very unhygienic. The Hotel was mentioned as dangerous to the health of migrants because one could easily contract gonorrhea and HIV/AIDS by patronizing sexual services from there. At the public toilet, due to the unhygienic nature of the place, migrants indicated the possibility of easily being infected with cholera and candidiasis ("white") for the women using the facility.

Areas like pharmacies, schools, churches, pharmaceutical company, and the police station were depicted to be safe by participants as far as their health is concerned. Churches were considered to be safe for the health of migrants due to the health education and "salvation" messages that will make one "to conduct himself/herself well to avoid illnesses." In other words, migrants perceive church messages as protective against risky health behaviors. Regarding the perceived link between the police station and health, migrants connected the police station to emotional and psychological health. For example, a participant remarked that "when one is attacked or threatened by another person, one could run to them for protection. Receiving constant death threat from your fellow human being could make you to develop BP [high blood pressure]." Hence, the existence of the police station gives migrants a sense of peace, and this could promote good health. Mention was made about educational campaigns embarked upon by some police officers in the neighborhood.

The pharmacies and drug store were perceived as safe and good for health because migrants reported to obtain medication from there when unwell. Migrants perceive the pharmacies and drug store as essential to access health care services whenever necessary; additionally, mention was made of community members receiving health education whenever they patronize these pharmacies and such education enable them to protect themselves from diseases or illness contraction.

Participatory Hazard Mapping Results, James Town: Males

Major landmarks in the community shown on the map by the male migrant participants included a drinking spot (a place of sale of alcoholic beverage), public toilet, churches, roundabout, police station, London market, an abattoir, a pharmaceutical company, and the fire service station (see Fig. 4.6). Dangerous places for health indicated on the map by migrants were the areas around the public toilet, London market, areas around one of the neighborhoods referred to as "Akoto Lantey," and the surroundings of an abattoir. Migrants remarked that the general unhygienic spaces make them dangerous for health. They reported that the abattoir and its immediate environment are so dirty with a lot of houseflies, which usually make people sick. The migrants indicated that people sometimes get diarrhea when meat from the abattoir is consumed. The areas around the London market and the public toilet were reported to have big gutters, which are choked and therefore breed mosquitoes and produce stenches thereby making community members get sick, especially those who live near those places. When probed, the three diseases mentioned by migrants to be because of these dangerous spaces are malaria, fever (other than malaria), and diarrhea.

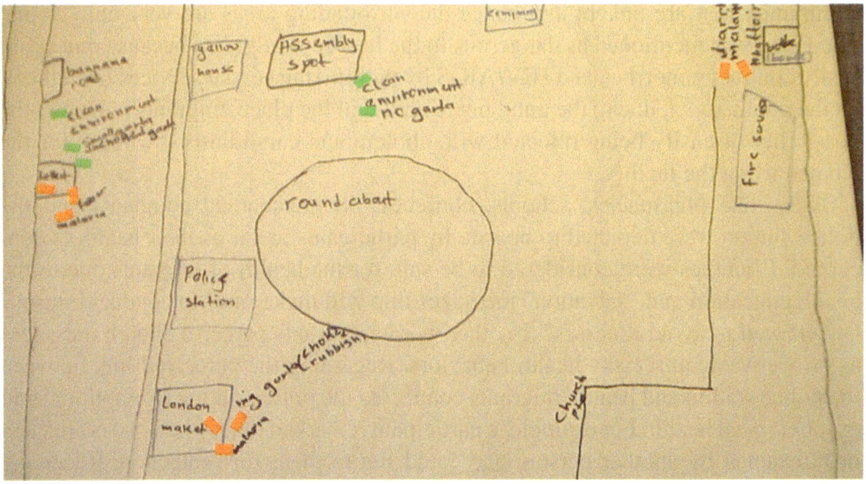

Fig. 4.6 Participatory hazard mapping, James Town (Males)

The male migrant participants indicated a few safe places for health. These were areas around a popular drinking spot and a sub-neighborhood. In the view of the study participants, areas such as assembly spot (drinking spot) and the vicinity of Banaman road were depicted as safe areas for health because such areas do not have open gutters. Existing gutters are also small such that community members are unable to put in garbage, which will cause these gutters to choke and consequently breed mosquitoes and produce a stench which will lead to the contraction of diseases such as malaria, fever (other than malaria), and diarrhea.

Participatory Hazard Mapping Results, Old Fadama: Females

The major landmarks in the community depicted on the map (sketch) include non-residential infrastructure such as the police station, drug stores, a market where wood is sold, cold store, church and mosque buildings, residential houses, hairdressing salon, grocery shop, public standpipe, the public bathhouse, and a private commercial toilet facility. Other landmarks reported were a community park commonly referred to as the Jima park, hair dressing saloon, a big drain (the Korle Lagoon gutter), and a subvicinity called Dr. Branch (Fig. 4.7).

Places such as the big drain called the Korle gutter, Dr. Branch vicinity, the public bathhouse, the mosque, the commercial toilet, the police station, and around the cold store were mapped up by participants as dangerous places for the health of migrants. The Korle gutter (a river) is a dangerous place for health because of the constant flooding it causes in the community any time it rains. The river has become a dumping site for some residents aggravating the flooding occurrence in the community.

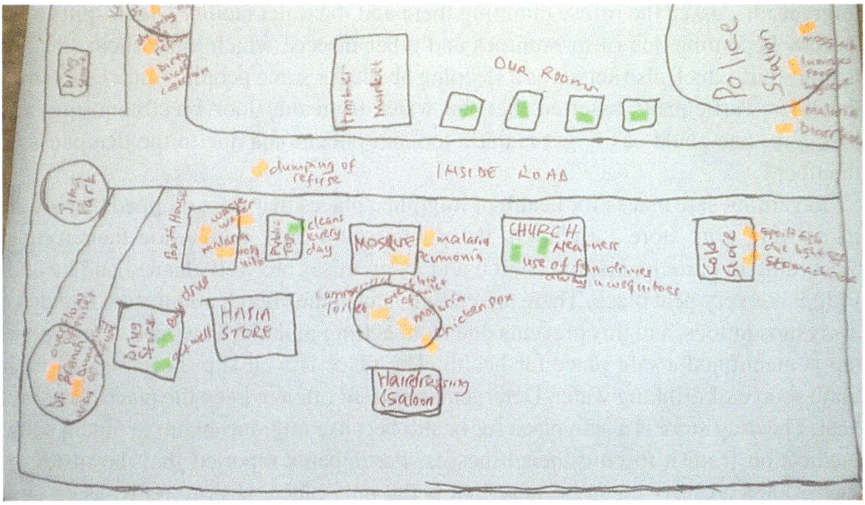

Fig. 4.7 Participatory hazard mapping, Old Fadama (Female)

So, some children have drowned when they played nearby the big gutter. The public bathhouse is also reported to be dangerous for health. Migrants indicated that due to the constant production of waste water, which becomes stagnant, it serves as breeding ground for insects and parasites thereby making people sick. Regarding the commercial private toilet, the place is not tidy and sometimes after using the place, one is not able to flush the feces. Therefore, participants asserted it also becomes a hub for mosquitoes and other insects causing diseases like malaria. The migrants observed that, sometimes, the septic tank overflows and that could spread other diseases such as cholera and chicken pox. About the cold store, during electricity fluctuations and loss of power, some of the fishes being sold go bad causing a heavy stench in the neighborhood. Migrants reported that sometimes when such fishes are sold to the community members, they end up getting stomach ache resulting from diarrhea. Migrants indicated that the immediate surroundings of the police station are very filthy. The place is not well kept. People dump rubbish and human excreta there. Migrants observed that some inmates who are detained at the police station defecate causing bad odor and the spread of illnesses, which they perceived as malaria and diarrhea. It was also noted that the main untarred road in the community is used as a refuse dump by some residents the moment it begins to rain because some community members prefer to dump their rubbish on the street so the rains "carry them away" instead of paying those informal refuse collectors called "kaya bola" to take them away. Other times too, migrants alleged that, "kaya bola" workers dump the refuse on the road, although they may have taken monies from their customers to send the garbage to the appropriate destination. It was noted that, on some occasions, a few of the "kaya bola" who were caught leaving refuse on the road were beaten and made to carry it away; a situation which migrants attributed to the fact that the community has no one in charge of sanitation issues. At Dr. Branch area, migrants discussed the refuse dumping there and the toilet facility which spills over thereby becoming den of mosquitoes and other insects, which make them sick. At the mosque, which also serves as a sleeping abode for some people, water leaks from the floor. Participants asserted that the water from the floor breeds mosquitoes; therefore, one could easily get malaria too and pneumonia due to the dampness of the floor.

Regarding safe places for health of migrants, places that were mapped by participants were drug stores, a church, the public stand pipe, and inside their rooms. According to participants, the church and its premises are well cleaned and has not filth. It is a very neat place. There are ceiling fans in the church auditorium that drive away mosquitoes, and this prevents one from getting malaria. The public stand pipe/ tap is mentioned a safe place for health. The place is well kept because it is their main source of drinking water. Detergents are used often to keep the place clean and neat. The drug store is a safe place for health because migrants claim to obtain some medication from it to cure their illnesses. Participants reported that the inside of their rooms are very clean because that is the only place they sit, relax, sleep, and eat. However, a few of them were quick to add that they are faced with the issue of stench from spilling over toilets, as well as stagnant and waste water.

Participatory Hazard Mapping Results, Old Fadama: Males

The major landmarks in the community depicted on the map (sketch) include non-residential infrastructure such as the police station, drug stores, cold store, church and mosque buildings, block molding factory, drinking bar (spot), a private commercial toilet facility, and the main untarred road within the community (see Fig. 4.8). Other landmarks reported were the community park commonly referred to as the Juma park, Dr. Branch (an old building that used to be a place managed by one medical officer as a community health post), and Obuasi area (most of the migrants used to be from Obuasi in the Ashanti region).

Places such as the Obuasi area, Dr. Branch vicinity, Juma park, a drinking bar (spot), and the commercial toilet were mapped up by participants as dangerous places for the health of the migrant. According to the participants, Dr. Branch vicinity is dangerous for migrants' health because there is always flooding when it rains. The vicinity has no drains to support the flow of rain water and therefore it becomes haven to mosquitoes and other insects. This, according to migrants, results in community members contracting diseases such as malaria, foot rot, typhoid, and jaundice. Migrants reported that when the rains have not begun, especially in the dry season, the entire area becomes so dusty resulting in air pollution thereby making residents suffer from cold and catarrh. People living in that vicinity engage in disposal of waste water in the open, and this could affect the health of the migrant. Also, the Obuasi area in the community was reported to have a lot of refuse in addition to the indiscriminate disposal of garbage in the community, which is not good for their health. Participants indicated that the private commercial toilet has a lot of

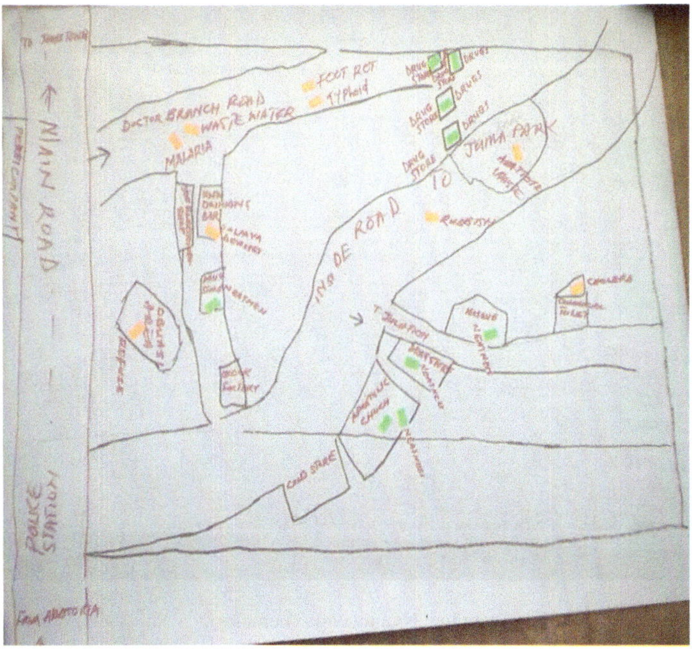

Fig. 4.8 Participatory hazard mapping, Old Fadama (Males)

houseflies and cockroaches and that could result in making residents contract cholera. The drinking bar (spot) is not very neat, and migrants find sputum all over the place, and this could spread diseases including cholera. With respect to the Juma park area, migrants reported that it is the main location for slaughtering of animals. The wastes from the animals are dumped right there; this leads to the place becoming a den of houseflies and other insects thereby spreading diseases.

About safe places for the health of migrants, the drug stores, the mosque, and church were mapped up by participants. The drug stores are considered safe because there are no houseflies in there. Operators of these stores, according to migrants, use air conditioners so the inside of these stores are very safe. One cannot contract any disease there "even if your food falls on the floor and you pick it back and consume it," said a participant. The mosque and the church too are very neat. These places are well kept, and it is safe for the health of any migrant. One will not contract any disease being there.

Participatory Hazard Mapping Results, Namuwongo: Females

Migrants drew a map of their community and singled out major landmarks as the market, the railway (railway line), police post, Nakivubo Channel, toilet, churches, and bathing shelter (See Fig. 4.9). Places they considered to be dangerous include

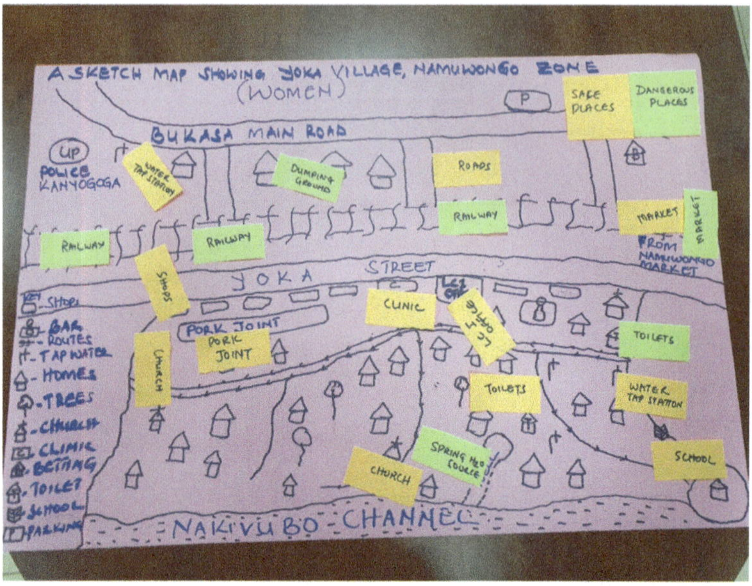

Fig. 4.9 Participatory hazard mapping, Namuwongo (Females)

the market that operates every day. They reported that the congestion of commercial motor cycles (boda boda) leads to accidents, traffic congestion, and loss of property to petty thieves. The railway line crosses through the community and was reported as one of the dangerous places because of its high speed and its inability to stop in case of an accident. This has led to many deaths of both children and animals. Migrants also reported that dumping grounds for waste disposal and toilets as very hazardous to health. They reported that there is poor disposal of fecal matter, whereby human waste is tied in polythene bags and dumped along other rubbish in rubbish dumps, leading to outbreak of diseases such as diarrhea, dysentery, and cholera. The spring water source was also reported to be a hazard to health because community members urinate in it and dispose of human and other waste in it. Children also use the same spring well as "swimming pool" leading to common diseases.

Churches, schools, water tap station, clinics, toilets, markets, Local Council offices, pork eating restaurants (commonly referred to as pork joints), shops, and roads were reported as safe spaces as far as health is concerned. Churches were singled out as safe because during times of seasonal flooding and when families have domestic conflicts, they are shelter community members. Church members also clean the community, have schools that teach children keeping them out of danger, and teach them good health habits. Churches were singled out as being a source of health-related messages to the communities. Clinics in the community also provide a safe place because they are source of information about health; they provide treatment for most of the diseases that migrants suffered from. The water tap station was also mentioned as a safe place because it is a source of clean water that is treated against germs, although it came at a cost. Toilets were also depicted to be safe places as regards health as they helped in the proper waste disposal in the community. Noteworthy is that these were public toilets that were paid for. They indicated that when one has diarrhea, they can use the toilet often, limiting the spread of germs and diseases to other members of the community. The Local Council offices were also mentioned as a safe place for community members as they mobilized the community members for communal work such as cleaning the environment around them and spread information about any emergencies in the community, such as disease outbreaks. They also bring health reinforcement teams that spearhead cleaning of the community.

Participatory Hazard Mapping Results, Nawumongo: Male

The male migrants drew a map of their community and singled out major landmarks such as the police post, Nakivubo channel, "pork joint" (where people gather to eat pork), the bathing shelter, the public toilet, sports betting rooms, nonofficial parking yards, and "drinking joints" (Fig. 4.10). Male migrants reported that some of the places they considered dangerous included the sports betting rooms, the market, the railway line, the police post, Nakivubo channel, and corridors between buildings, which harbor idlers. The sports betting rooms were said to be dangerous because many youths spend time gambling without doing any work that benefits their lives.

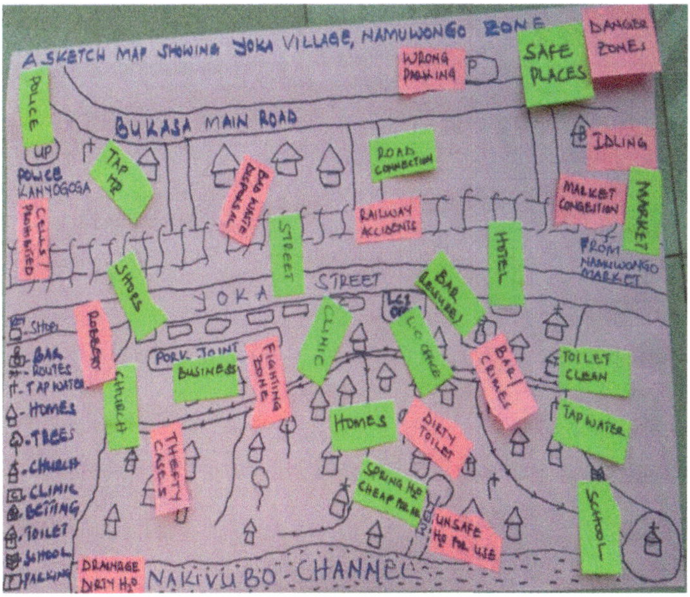

Fig. 4.10 Participatory hazard mapping, Namuwongo (males)

The police post was singled out as both a good place because they arrest and detain criminals, and a dangerous place because sometimes they arrest innocent people and have them detained.

The male migrants reported safe spaces in the community as the police post because they arrest criminals especially those selling narcotic drugs in the community. The "pork joint" where people eat pork and consume alcohol was also singled out as a safe space, where men gathered and bought each other drinks and shared pork together. They argued that while drinking alcohol they tend to solve a lot of domestic issues. In addition, the worshipping places, schools, clinics, and toilets were also mentioned as some of the safe spaces, especially as regards health.

Participatory Hazard Mapping Results, Kinawataka: Females

The major landmarks identified in this community included schools, police barracks, water trenches, water springs, churches, mosque, roads, the railway line, dumping zones, the forest, leisure centers, and clinics/drug shops (Fig. 4.11). Residents reported that the trenches were the most hazardous places in the community because they were used as ungazetted dumping places, and they are the abodes of human waste and waste from abortions from community members. The trenches have no bridges and yet they have water flowing across them. Many children were reported to have fallen in the dirty water in the trenches and end up galloping the dirt water. Some of the homes were so close to the trench that their doors opened into the trench. The trenches were also the conduits for effluents from the Nomi soap

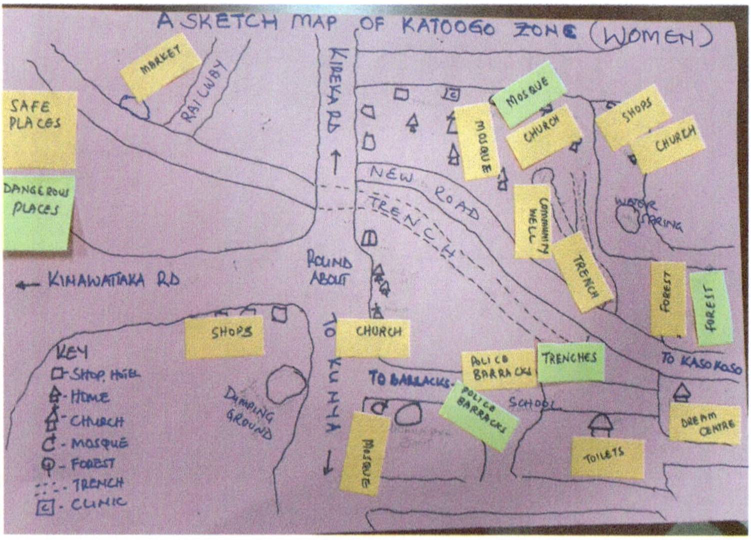

Fig. 4.11 Participatory hazard mapping, Kinawataka (female)

factory making it dangerous to the community members. Migrant residents reported that during the rainy season, people open their septic tanks containing human waste to flow into the trenches because they have no other way of emptying their toilets. In addition to all these dangers, the trenches often flood during the rainy season bringing all these components contained in flood into their homes, as they also tend to flood since it is a wetland. The local name for this wetland is "Katoogo" meaning swamp; it was a swamp but now claimed by migrants for residence.

The police barracks, a residential place for police officers, was also reported to be hazardous to the community because they release their human waste into the community well, yet there is a water shortage. The police in the barracks dump their garbage on the road side and the place smells. The dumping grounds around the community were also reported as hazardous due to the stench that emanates from them and due to fact that they are not designated as such; they tend to fill up without a plan for emptying them. The mosque was also reported as a dangerous place because many people around them dispose off their rubbish anywhere in the community. This was in addition to the nearby forest that many drug users have turned into a haven for their illicit activities and the washing bay where people tend to fall in the trench nearby.

Migrants reported some safe places in the community such as toilets, the Dream Centre, the police barracks, churches and mosques, community well, shops, and the market. The toilets were singled out as good for proper waste disposal; however, it was reported that there are a few toilets compared to the number of migrants who use them. The Dream Centre, a women empowerment space, was singled out as being a safe place because women freely meet there for various purposes and it also has a Savings and Cooperative Societies (SACCO) that economically empowers the women. The Barracks in Kinawataka was also reported to be a safe place because of ensuring security, and it also houses a public primary school where children go to learn. The churches and mosques were also perceived as safe places because people visit them for spiritual development and renewal. Attendance to church and mosque were perceived as reducing people's free time, during which they might engage in useful activities to the whole community. The community well, shops, and the market were also reported to be safe places.

Participatory Hazard Mapping Results, Kinawataka: Males

The major landmarks according to the male migrants include roads, the railway line, trenches, schools, churches and mosques, a forest, police barracks, bars, Donald pork joint, washing bay and dumping zones, and clinics/drug shops (Fig. 4.12). Migrants selected the police barracks as a hazardous place because the residents dump waste by the road side and that cause the neighborhood to smell. The "forest" area was also reported as hazardous because women and young girls are raped, and robbers use it as a hiding place. The male migrants also reported that trenches are dangerous because people dump fecal matter and garbage during the rainy season.

Fig. 4.12 Participatory hazard mapping, Kinawataka (males)

Due to the flooding, this causes feces to flow in running water and causing numerous diseases to the community members. They claimed that the trench is so small and cannot contain the flow most of the times causing it to overflow into houses and nearby buildings. The mosque in the community was mentioned among the unsafe places because many people get strangled around it and most of those who stay nearby are very dirty. They reported that there are no nearby toilets for people in this area to use, to the extent that some use basins and polythene bags as toilets. The area around the mosques was said to be the dirtiest place in the whole community. Kyaks bar was also reported as dangerous due to the noise pollution it emitted in the community in addition to the prostitution and drug abuse that happen there. In Kyaks bar, even young children were allowed to sit and consume alcohol as adults. Many people got stabbed due to senseless fights in the bar. Due to prostitution, many children were caught using used condoms as balloons that led to catching diseases such as diarrhea and continuous cough.

Bars, police barracks, toilets, roundabout, market, and shops were reported to be some of the safe places in the community. Bars, migrants reported that they provide happiness and relieve residents of stress. For men, it was a place where they socialized with their friends and other community members. At bars, the males also reported that they receive condoms from organizations. The police barracks were safe in that they provide security for the community for both property and people. The shops provided easy access to household appliances and other items that were bought and used daily. It was also a source of employment for the community members. The market was reported as a safe space for providing a source of livelihood for many residents since most of the community members are venders in the market. On the whole, migrants indicated that most of the safe places were seasonally safe due to the flooding in the rainy seasons that affects most of the households. When the area floods, most of the houses and buildings are filled with water and there is no safe place to go.

Brief Deliberation on Participatory Hazard Mapping

The participatory hazard mapping activities revealed migrants' perceptions of their places of residence and environmental risks to health. Specifically, place characteristics and its relationship with perceptions of environmental risk to health were apparent. For example, the presence of the unkept surroundings of the abattoir in James Town determined the kinds of risks and associated diseases, whereas the soap factory in Kinawataka influenced perception of risks in that place of residence for migrants. However, we observe that migrants demonstrated limited knowledge of diseases that may be associated with the environmental hazards in their neighborhoods. While migrants could make linkages between filthy drains and foodborne diseases such as diarrhea and cholera, there was lack of comprehension on the origins of malarial vectors. More so, there was limited knowledge of the chemical and biological contaminants in the communities since they did not appear to be on the radar of migrants as they discussed dangerous places for health. Although some groups rightly identified an environmental risk to health, they were unable to specify plausible diseases that may be associated with such risks.

Worthy of note is the mischaracterization and misperception of dangerous places and the diseases that may be associated with them. For example, while the inadequate drainage system and its attendant flooding may result in ponds that may breed insects, male migrants in Old Fadama concurred with each other that it leads to jaundice. Similarly, males in Kinawataka attributed persistent coughing and diarrhea by children to blowing air into used condoms not properly disposed of.

We notice a lot of similarities and subtle differences in between the male and female groups regarding their perceptions of risks and their relations to diseases across all the places. While most area identified as dangerous to health and safe for health was almost uniform across the sexes, there were differences in perceptions of place regarding its status as dangerous or otherwise for health. For example, while bars were mentioned as safe for health because it has positive emotional influence among the men in Kinawataka, the women thought otherwise. The need for conscious education that focuses on specific dangerous places for health with exact emphasis on all possible negative health ramifications may be the way to enhance health literacy and enhance knowledge beyond the usual infectious diseases that are known to residents of poor neighborhoods.

Food Safety and the Environment

Slums have been known to experience food safety problems due to both exogenous and personal and household variables. Foodborne diseases occur frequently in informal settlements and slums in general compared to formal settlements and upscale neighborhoods (Tutu et al. 2019; Redmond and Griffith 2003). Therefore, given the exposure to varied environmental contaminants due to poor environmental management, perceptions on food safety and foodborne diseases were explored. We sought to examine the relationship between knowledge on food safety and receipt of food safety information.

Using a survey questionnaire, participants were asked how common they thought it was for people to contract diseases because of food; whether they thought contamination of food by germs is a food safety problem in their community (serious problem—1, somewhat a problem—2, not a problem—3). Information pertaining to possible unsafe food as perceived by participants were elicited, as well as a proxy for healthy food preparation practices (how often they washed their hands with soap before beginning food preparation). Knowledge on bacteria and food-poison causing agents was obtained. Specifically, migrants were asked if they had heard of Salmonella, Listeria, Campylobacter, *E. coli*, and Toxoplasma. Participants were asked how likely they thought raw foods contain germs or other microorganisms that could make you sick (not at all likely—1 to very likely—5). These foods are raw beef, raw fruit, raw vegetable, raw chicken, raw fish, raw shellfish, and raw milk. Additionally, migrants were asked if they have ever received information regarding food safety and/or foodborne diseases (yes—1, no—2), as well as perceptions on the link between hand washing and getting sick. Specifically, we asked: if you forget to wash your hands before you begin cooking, how likely are you to get sick?

Analyses

The analyses of the survey data on food safety consisted of descriptive statistics, computation of composite scores, bivariate analyses, and multivariate analyses. Descriptive statistics show distribution of responses to migrants' perception on foodborne diseases commonness, knowledge on possible unsafe food, and knowledge on food-poison causing organisms. Three health literacy composite scores were computed: (1) health literacy on possible unsafe food score; (2) knowledge on food-poison agents score ("food bacteria literacy"); and (3) health literacy on germs in raw food score ("food contains germs score"). Regarding bivariate analyses, chi-square test and independent samples *t*-test were used. Chi-square test was used to assess the association between the discrete demographic characteristics of the sampled migrants (e.g., sex, marital status, employment status, occupation, ethnicity, and religious affiliation) and two outcome variables: (a) ever received information regarding food safety, and (b) perception of food safety as a problem in the community (serious problem—1, somewhat—2, not a problem—3). Independent samples *t*-test was used to examine bivariate associations between demographic variables (continuous) such as age, age at migration, and household size, as well as the health literacy on possible unsafe food score, knowledge on food-poison agents score, and health literacy on germs in raw food score, and ever received information regarding food safety. ANOVA was used to test the association between the demographic continuous variables, the health literacy scores, and perception of food safety as a problem in the community. Binary logistic regression was used to examine the relative effects of the health literacy scores on receipt of food safety information controlling for demographic characteristics. Ordinal logistic regression was used to assess the relative effects of the health literacy scores on perceptions on food safety as a community problem controlling for individual characteristics.

Illustrations from Old Fadama and James Town

More of Old Fadama participants (75%) reported that they thought it was common to contract diseases because of food than James Town participants (64%) although both were of high proportions (see Table 4.3). Regarding food safety as a problem in the community, while 70% of participants in James Town reported that it is serious or somewhat of a problem, 85% of respondents in Old Fadama indicated same. While the numbers are high in both slums, they are substantially higher in Old Fadama. Foodborne diseases are commonly reported in James Town and it is known to be cholera endemic (Tutu et al. 2019). That may account for the perception of migrants regarding the severity of the health concern. Although studies on foodborne diseases in Old Fadama is scarce, we speculate that the perception of migrants about the seriousness of foodborne diseases is tied to the poor environment that makes food contamination easy to occur. Also, as discussed in Chap. 3, enteric fever being among the top 5 most reported diseases in the clinic in Old Fadama coupled with the mention of typhoid fever during the body health mapping, it complements the findings on migrants' perception about the seriousness of foodborne diseases in the slum.

Poor knowledge on possible unsafe food seem to be a major concern. In Old Fadama, while 33% reported that unwashed fruits may be possible unsafe food,

Table 4.3 Perceptions on commonness and severity of foodborne diseases in James Town and Old Fadama

	James Town		Old Fadama	
	N	%	*N*	%
How common				
Very common	53	35.3	63	42.0
Somewhat common	20	13.3	25	16.7
Common	23	15.3	25	16.7
Not common	40	26.7	20	13.3
Don't know	14	9.3	17	11.3
Severity or seriousness				
Serious problem	56	37.3	78	52.0
Somewhat	48	32.0	50	33.3
Not a problem	22	14.7	14	9.3
Don't know	24	16.0	8	5.3
Total	150	100.0	150	100.0

74% indicated that raw milk is not a possible unsafe food. Although 53% reported that undercooked poultry may be unsafe, 72% acknowledged that uncooked food may be unsafe food. Similar trends were observed in James Town. Twenty-one percent of migrant participants reported that unwashed fruits may be possible unsafe food, whereas 67% said that raw milk is not a possible unsafe food. While 51% indicated that undercooked poultry may be unsafe, 26% thought that uncooked food is not a possible unsafe food.

Regarding bacteria and food-poison causing agents, the poor knowledge among migrants were staggering. In James Town, less than 5% reported to have ever heard of Salmonella, Listeria, Campylobacter, *E. coli*, and Toxoplasma compared to less than 2% of migrants sampled in Old Fadama.

From the bivariate analyses, in Old Fadama, only marital status was significantly associated with receipt of food safety information and no statistical significance was observed for James Town. Regarding the perception of food safety as a problem in the community, in Old Fadama, the variables that were found to be statistically significantly associated with it were educational level, hygiene practices (frequency of washing hands), and perception on the link between hand washing and getting sick. The perception of food safety as a problem in the community did not yield statistical significance with any predictor variable.

Thus, an ordinal logistic regression model was performed for Old Fadama (Table 4.4). The model diagnostics (model fitting information and test of parallel lines) suggests that the data fits the model well. Therefore, the assumption undergirding the proportional odds statistics have not been violated. Those with no education in Old Fadama (coefficient 1.423; $p < 0.005$) had poorer rating on the outcome

Table 4.4 Ordinal logistic regression of food contamination in Old Fadama

		N	Ordered log odds
Contaminated food a problem in your area	Serious problem (1)	78	−1.703
	Somewhat a problem (2)	50	0.538
	Not a problem (3)	22	
Education	No education	47	1.423*
	Primary	28	0.728
	Junior high school	49	0.591
	Senior high school and higher	26	
Often wash hand before food preparation	All the time	51	−2.155*
	Most of the time	42	−1.115
	Some of the time	27	−0.47
	Rarely	17	−1.36
	Never/don't know	13	
Perception on handing and getting sick	Very likely	79	−0.802
	Likely	40	−1.05**
	Quite likely	31	

*$P < 0.005$; **$P < 0.05$; Nagelkerke $R = 0.342$

variable. That is, they were less likely to perceive food contamination in Old Fadama as a serious problem compared to those with senior high school or higher education. Migrants who reported to wash their hands all the time with soap before food preparation had better rating on the response variable (coefficient −2.155; $p < 0.005$). Those migrants were more likely to perceive food contamination as a serious problem in Old Fadama compared to the migrants who reported to never wash their hands with soap before food preparation.

Brief Deliberation on Food Safety and the Environment

Firstly, high proportions of residents in the communities recognize that food safety and contaminated food is serious or somewhat of a problem. While the survey does not probe for specific reasons for which respondents think of food contamination as a problem, we speculate that the results of the body health mapping in Old Fadama, the perennial cholera outbreak in James Town, and the number of reported typhoid fever cases in Old Fadama may be the contributing factors feeding that perception. Additionally, in both slums, references made by participatory hazard mapping participants to the dangers posed by the abattoirs in their community may be contributing to the recognition of food safety challenges. See Figs. 4.13 and 4.14 for pictures of the abattoir in James Town to which the participatory hazard mapping respondents referred.

Secondly, there seem to be poor knowledge on unsafe food. Although a lot of migrants surveyed believe food may be unsafe, high percentage did not think certain food may be unsafe. For example, a large proportion did not think that raw milk may

Fig. 4.13 Pictures of the abattoir in James Town (authors' fieldwork)

Fig. 4.14 Pictures of the abattoir in James Town (authors' fieldwork)

be unsafe. Also, in James Town, 21% of respondents did not perceive unwashed fruit as a potential health hazard. Additionally, there was overwhelming knowledge deficit on bacteria contaminants that are known for diverse forms of food poisoning. While we recognize that most of them may not be household names, the low knowledge level of *E. coli* was astonishing.

Thirdly, the perception about the problematic nature of contaminated food in Old Fadama is a function of education and personal hygiene. Migrants with higher education were more likely to perceive contaminated food as a problem in the community. Respondents who were in the habit of washing their hands with soap and water before food preparation were more likely to think that contaminated food in the community is a serious problem. These findings make the suggestions for actions to be taken to decrease environment-related disease burden more relevant. Actions such as use of environmental contaminants home improved hygiene and household water consumption measures, as well as executing best sanitation protocols.

References

Briggs DJ (2008) A framework for integrated environmental health impact assessment of systemic risks. Environ Health 7(1):61

Brown R, Leder K, Wong T, French M, Chown SL, Luby S, Clasen T, Reidpath D, El Sioufi M, McCarthy D, Forbes A (2018) Improving human and environmental health in urban informal settlements: the Revitalising Informal Settlements and their Environments (RISE) programme. Lancet Planet Health 2:S29

Ezzati M, Utzinger J, Cairncross S, Cohen AJ, Singer BH (2005) Environmental risks in the developing world: exposure indicators for evaluating interventions, programmes, and policies. J Epidemiol Community Health 59(1):15–22

Fazzo L, Minichilli F, Santoro M, Ceccarini A, Della Seta M, Bianchi F, Comba P, Martuzzi M (2017) Hazardous waste and health impact: a systematic review of the scientific literature. Environ Health 16(1):107

Filcak R (2010) Migration to contaminated sites: migrants' settlements in Central and Eastern Europe built in places with high environmental and social vulnerability. *Environment, Forced Migration And Social Vulnerability*, pp 133–143. Springer, Berlin, Heidelberg

Prüss-Üstün A, Corvalán C (2006) Preventing disease through healthy environments: towards an estimate of the environmental burden of disease, vol 12. World Health Organization, Geneva

Redmond EC, Griffith CJ (2003) Consumer food handling in the home: a review of food safety studies. J Food Prot 66:130–161. https://doi.org/10.4315/0362-028x-66.1.130

Tutu RA, Gupta S, Busingye JD (2019) Examining health literacy on cholera in an endemic community in Accra, Ghana: a cross-sectional study. Trop Med Health 47(1):31

Vilain J, Galliot AM, Durand-Roger J, Leboyer M, Llorca PM, Schurhoff F, Szoke A (2013) Environmental risk factors for schizophrenia: a review. *Encephale* 39(1):19–28. https://doi.org/10.1016/j.encep.2011.12.007

World Health Organization (2008) Health environment: managing the linkages for sustainable development: a toolkit for decision-makers: synthesis report. WHO, Geneva

World Health Organization (2016) Preventing disease through healthy environments: a global assessment of the burden of disease from environmental risks. WHO, Geneva

World Health Organization (2017) Drinking-water fact sheet. http://www.who.int/mediacentre/factsheets/fs391/en. Accessed 20 Jan 2018

Chapter 5
Health Literacy of Migrants: Environmental Risks to Health

Abstract This chapter focuses on the health literacy of migrants regarding environmental risks to health. Health literacy as a concept is first interrogated and conceptualized, positioning it as both a conceptual framework and measuring tool. Assessing health literacy was achieved by using survey questionnaires and conceptual mapping to demonstrate people's perceptions of environmental pollution and how information was accessed about environmental contamination. Basing on the WHO recommendation of measuring health literacy in developing countries, the same parameters were used to measure health literacy in the participating communities. Results from the survey revealed that the migrants were aware of environmental pollution and they had access to information about it. Findings further revealed poor *intentionality* in seeking health information although general literacy impacted health literacy positively. The conceptual maps revealed that fecal contamination in the communities was perceived to be directly because of inadequate toilet facilities as well as unhygienic practices, which results in the spread of diseases. Most importantly, we notice big gaps in disease knowledge. Although mention is made of diseases such as cholera, dysentery, and typhoid, other predominant diseases associated with fecal contamination such as hepatitis, ascariasis, and gastroenteritis are not stated.

Keywords Health literacy · Conceptual mapping · Mental models · Fecal contamination · Disease knowledge · Foodborne diseases · Urban poverty · Slum sanitation

Introduction

Health literacy is a vital component of health geography and of relevance to health outcomes. In this chapter, we begin by examining the multifaceted nature of the concept of health literacy. We digest its multiple constructs and diverse contexts across scale. A brief discussion on major tools that have been developed to measure

R. A. Tutu, J. D. Busingye, *Migration, Social Capital, and Health*, Global Perspectives on Health Geography,
https://doi.org/10.1007/978-3-030-24693-8_5

health literacy is undertaken. Subsequently, the health literacy of migrants on environmental risks to health, which was discussed in Chap. 4, is assessed. Using conceptual mapping (mental models), we explore migrants' perception on the causes and consequences of fecal contamination in the communities.

Health Literacy: A Conceptual and Practical Tool

The concept of health literacy have now been comprehended to be a critical component of social determinants of health, which affects the health outcomes of people in varied contexts with varying forms of health conditions (DeWalt et al. 2004; Schillinger et al. 2002). Although the concept has been defined differently with emphasis of varying components, the World Health Organization (WHO) defines health literacy as "the personal characteristics and social resources needed for individuals and communities to access, understand, appraise and use information and services to make decisions about health" (WHO 2015). That is, health decision-making is inextricably connected to individual and societal characteristics. For example, health-related decisions may be made by an individual for him/herself, a family member, the entire family, or a community at large. And since the decisions made are dependent on the health literacy of that individual, it has implications for the health behaviors of the other family members as well as the community.

The multiplicity of health literacy conceptions has led to diverse conceptualization and application in several domains. We discuss earlier conceptualizations on health literacy and show how they have been applied in various contexts (Nutbeam 2000; Institute of Medicine 2004; WHO 2015). Nutbeam's (2000) developed a health promotion-oriented model. He identified three health literacy dimensions namely functional health literacy, interactive or communicative health literacy, and critical health literacy. These typologies, Nutbeam asserts, are linked precursors embedded in health promotion actions, which he calls "content," such as advocacy and education. According to him, functional health literacy is about the individual having enough basic writing and reading capacity to function effectually in everyday circumstance. Interactive health literacy refers to a combination of more advanced form of mental and general literacy skills as well as social skills that can be used to obtain information and deduce meaning from diverse communication forms and utilizing it to varying life conditions. Critical health literacy implies more advanced form of mental and general literacy skills, together with social skills allows an individual to critically study health information and use that information to exercise control over their life's changing circumstances. Later, Nutbeam (2008) intimates that conceptualizing health literacy in that manner has vital implications for changing peoples' lifestyles through enabling comprehension of prescription medication, improving peoples' knowledge and their competence to act. That sentiment is shared by Sorensen et al. (2012, pp. 4–5) who assert that Nutbeam's conception of health literacy "represent levels of knowledge and skills that progressively support greater autonomy and personal empowerment in health-related decision-making, as well as engagement with a wider range of health knowledge that extends from personal health management to the social determinants of health."

Conceptually, Nutbeam's typologies have expected positive outcomes at the individual and social scales given their respective content. On functional literacy, the communication of factual health information on health risks and services use may lead to improved health knowledge on health risks and health services utilization at the individual scale while leading to increased involvement in population health program such as immunization at the community scale. The content for interactive health literacy, in addition to that from the content of functional health literacy, will involve openings for the development of skills within a helpful surrounding. At the individual scale, it is expected to result in improved capability to act independently on health knowledge. At the community scale, it will lead to improved ability to affect social practices. Regarding critical health literacy, added to the content of functional health literacy and communicative health literacy, providing information on socioeconomic determinants of health may lead to improved individual ability to withstand socioeconomic stressor, while at the community scale it is expected to result in enhanced ability to act on social determinants of health and empower the community.

The Institute of Medicine framework for health literacy is based on two primary positions: (1) that literacy is at the basis of health literacy, and (2) health literacy is the mediator between individuals and their respective health contexts (Institute of Medicine 2004). In other words, individual capabilities are the major premise for how health literacy will impact health outcomes. Thus, while literacy is the skills conduit through which individuals comprehend and communicate health issues, there are factors such as mental capabilities, bodily states, and social abilities that individuals convey to health situations. Therefore, literacy influences health literacy as a mediator between individual factors and health context to affect health outcomes. The relationship between health literacy and actions in health context implies the need for individuals to possess certain health literacy skills for effective performance in varied health context. These skills have been illustrated by the Institute of Medicine using a several examples including, but not limited to, find the way to and around the healthcare system, active participation during interaction with healthcare experts, comprehension of individual consent, as well as the application of health information through one's life course and events. Regarding finding one's way within the healthcare system, skills required consist of completing health application and insurance forms, comprehension of printed patients' rights and obligations, as well as actual way finding within a complex hospital or clinic setting. Concerning active participation during interaction with healthcare professionals, skills identified include seeking explanations, asking pertinent questions, arriving at knowledgeable decisions based on information received through such interactions, as well as working in partnership with healthcare provider to create a plan for the management of diseases. With respect to comprehension of individual consent, the person should be able to understand the content of informed consent forms for both research and medical procedure. On application of health information through one's life course, important skills will comprise reading and utilizing health information involving childcare and eldercare, reading and interpreting safety precautionary measures at work, choosing a best-fit healthcare plan, and determining and adopting strategies for augmented physical activity at an older age.

Baker (2006) supplemented the Institute of Medicine's model (2004) through an inference from the concept of the individual *dynamic state*—the idea that effective mediation by health literacy is not only dependent on the characteristics of the individual but also the healthcare system—by suggesting domains of health literacy and the relationship between health literacy (print and oral) and health outcomes. In Baker's model, he stresses two main individual capacity: (1) reading fluency—ability to read and decipher text, capacity to apply basic arithmetic, and capacity to use materials in health document, and (2) prior knowledge—how the individual conceives and knows about the healthcare system as well as how vest he/she is on general vocabulary.

The model intimates that reading fluency influences ability to comprehend written health information (health-related print literacy) and prior knowledge influences both capabilities to understand written health information and ability to interact verbally about health (health-related oral literacy). Subsequently, both health-related print literacy and health-related oral literacy impact the gaining of new knowledge by the individual, promoting positive attitudes, enhancing self-efficacy, as well as leading to positive behavior change thereby leading to improved health outcomes. It is worth noting that Baker makes a compelling case about the need to distinguish general literacy abilities from health literacy capabilities. Although the ability of an individual to read and write is likely to influence ability to read and write health information, it is crucial to make a distinction during research because ability to read and appreciate health information is more likely to impact health outcome than general literacy. Subsequently, the role of health literacy as a mediator between individual background characteristics and outcomes (health status and preventive health behavior) was test by Bennett et al. (2009). In a study on the contribution of health literacy to health status disparities and preventive health behaviors among older adults, health literacy was a significant mediator for both ethnic and education-related disparities in health status as well as influenza vaccination receipt.

A varied conceptual model of health literacy have considered both individual scale factors as well as contextual scale factors that affect an individual's health literacy with consequences for health outcomes. Paasche-Orlow and Wolf (2007), in theorizing and establishing the causal paths linking health literacy and health outcomes, concluded that health literacy should be perceived as both individual and healthcare system phenomena. Individual patient sociodemographic factors such as ethnicity, education, age, occupation, income, employment, social capital, and culture are identified to affect health literacy. Individual physical characteristics that impact health literacy in their model are vision, hearing conversational abilities, memory, and reasoning. Health literacy subsequently impacts individual and context factors in three domains—access and utilization of health care, provider–patient interactions, and self-care. The framework intimates that access and utilization of health care is predicated on (1) patient factors such as navigation skills, personal abilities, and perception on impediments; and (2) system factors—complexity, acute care proclivity, and tiered delivery model. Patient factors in the context of provider–patient interaction are knowledge, viewpoints, and involvement in decision-making, whereas the provider (system) factors in the same context include interaction skills, teaching capability, time, and patient-oriented care giving. Regarding self-care, the authors indicated personal motivation, skills, self-efficacy, and problem-solving skills as patient factors, while extrinsic factors noted are the presence and influence of mass media, the availability and impact of health education, and support technologies. Therefore, while access and utilization of health care affects provider–patient interaction and self-care, they interact to influence health outcomes.

In an effort to reduce ambiguity in analyzing and discussing health literacy in nursing research and practice, Speros (2005) presented a conceptual analysis of the concept. Speros asserts that "health literacy should become a variable assessed by nurses at the initiation of any client encounter, as is mental status, past medical history, and socio-economic status, so that recommendations and the plan of care can be adapted accordingly" (p. 638). She presented a set of attributes, antecedents, and consequences of health literacy. The primary attributes of health literacy she identified are reading capacities, comprehension, numeracy proficiencies, and ability to apply health information to make health-related decisions and functioning successfully as a patient in a healthcare setting. The precursors for those attributes, she argues, are twofold: (1) literacy—comprising capability to read, understand written words, and level of proficiency in numeracy comprehension; and (2) health-related experiences consisting of familiarity with medical terminologies and logical context within cognitive framework. The consequences for positive attributes and antecedents, according to Speros, will include improved self-rated health status, reduced use of healthcare services, reduced cost of health care, improved health knowledge, as well as shorter duration of hospitalization.

To understand (1) the causal pathways or intermediary stages that connect low health literacy to poor health status and high utilization of costly healthcare services and (2) the effects of social support on the associations between health literacy and health service use, Lee et al. (2004) developed a model that distinguished four dimensions of significance to health status. These dimensions or intermediate factors link health literacy to three health outcomes namely hospitalization, emergency care, and health status. The intermediate factors are knowledge of disease and self-care efficacy, health risk behavior, preventive care and physician visits, and medication compliance. The authors indicate that people with lower health literacy are more likely to engage in unhealthier behavior, nonadherence to preventive care measures, poor medical knowledge, and poorer compliance with medication. As a result, these factors lead to negative health outcomes including increased and frequent use of emergency care and high hospitalization. Lee et al. (2004) acknowledge that the effects of health literacy on the identified health outcomes are predicated on the influences of antecedents such as sex, ethnicity and culture, health insurance coverage, income disparities, and other socioeconomic status variables. The pathways proposed, according to Lee et al. (2004), drew on extensive previous studies. With respect to disease knowledge and self-care, existing literature pointed to evidence that individuals with poor health literacy had poorer disease knowledge. On preventive care, the model intimated that one's inability to comprehend health information about the implications and consequences of early detection of disease may lead to poor utilization of preventive care. Regarding health risk behavior, although they hypothesized that individuals with lower health literacy may be more likely to engage in lifestyles that are detrimental to good health like substance use and smoking, recent studies have confirmed their hypothesis (Suka et al. 2015; Liu et al. 2015). Therefore, Lee et al.'s (2004, p. 1313) first proposition was "*Other things being equal, individuals with lower health literacy will have poorer health status and greater utilization of hospital and emergency care services as mediated by their lower disease and self-care knowledge, worse health behavior, more limited use of preventive and physician care, and lower compliance with medication.*"

Relating to the effects of social support on the associations between health literacy and health service use, Lee et al.'s propositions are premised on the inextricable link between individuals and their societies. Individuals are social beings who live in social surroundings containing diverse amounts of resources and support. Therefore, to appreciate the extent which low health literacy may lead to poor health outcomes, it is crucial to control for the resources individuals gain through social support systems when stressors arise, including problems encountered due to poor health literacy. The authors identify two main social support moderators: structural social support and functional social support.

Structural social support refers to an individual's position in the social structure. For example, involvement in community and civic organizations, continuation and preservation of social contacts, and maintaining close knit social relations. One's positions within the structure may facilitate the possibility of gaining resources that may enable buffering against life stressors including adverse health outcomes. In the context of health literacy, structural social support acts as a buffer against adverse consequences of poor health literacy through information sharing, especially in close knit social networks—family and friends are usually the first people who are consulted regarding health issues. Additionally, apart from the sense of belonging that may enable people to face stressful circumstance thereby enhancing efforts to cope with low health literacy, structural social support may counteract the negative impact of low health literacy by compelling the individual to conform to healthy practices of a group. Consequently, their second proposition is "*Other things being equal, the adverse effects of low health literacy on health status and utilization of hospital and emergency care services will be greater among people with less structural social support compared with those who have greater structural social support. Furthermore, such interactive effects are mediated through disease and self-care knowledge, health risk behavior, use of preventive and routine physician care, and compliance with medications*" (Lee et al. 2004, p. 1315).

Functional social support comprises communication and reciprocal endeavors that serve a range of emotional and physical needs. Emotional support from social relations encourages feelings of self-worth that assist people to cope with individual inadequacies including poor health literacy and its attendant adverse health impacts. Tangible support from social relations in the form of aiding the navigation of complex health systems and communicating with health professionals may reduce the anxiety associated with circumnavigating the healthcare system by an individual with low health literacy. Therefore, their next proposition was "*Other things being equal, the adverse effects of low health literacy on health status and utilization of hospital and emergency care services will be greater among people with less functional social support compared with those who have greater functional social support. Furthermore, such interactive effects are mediated through disease and self-care knowledge, health risk behavior, use of preventive and routine physician care, and compliance with medications*" (Lee et al. 2004, p. 1315).

The multidimensionality of health literacy has led to the development of several tools for its measurement. Key among such tools include Rapid Estimate of Adult Literacy in Medicine (REALM-R) (Murphy et al. 1993), the Test of Functional Health Literacy in Adults (TOFHLA) (Parker et al. 1995), Health Activities Literacy Scale (Educational Testing Service 2006), and the Newest Vital Signs (Weiss et al. 2005). REALM, a 66-item instrument, was developed to measure various domains of vocabulary through word recognition and pronunciations. TOFHLA on the other hand was developed to assess the degree of reading comprehension and reading fluency of individuals regarding hospital documents and prescription labels. While the Health Activities Literacy Scale measured individual's comprehension of text and quantitative abilities in the areas such as health promotion, disease prevention, and health protection, the Newest Vital Signs focused on documents and quantitative information skills test. A lot of criticisms have labeled against these instruments. These include how long they take to administer, not suitable for most cultural settings, as well as the discomfort associated with asking research participants to read text aloud (Bann et al. 2012; Wolf et al. 2007).

New waves of instruments have been subsequently developed for specific diseases and context. Others have been developed to match specific health literacy definitions (e.g., Bann et al. 2012; Chinn and McCarthy 2013; Chung and Nahm 2015; Kim et al. 2012). In order to operationalize a modified Ratzan et al.'s (2000) definition of health literacy that consisted of the extent to which an individual can acquire, process, comprehend, and communicate concerning health information required to make informed health decisions, Bann et al. (2012) developed a health literacy instrument that measured print literacy, numeracy, oral literacy, and ability to seek information using the internet. While Chung and Nahm (2015) developed the eHealth Literacy Scale in the context of eHealth technologies to meet the increasing use of cyber space in a digital age, Chinn and McCarthy (2013) on the other hand, having identified the limitations in the previous instruments, filled these gaps by developing an instrument that focused on comprehending health literacy in social and institutional context including among healthcare providers.

More recently, the WHO (2015) has suggested seven considerations for studying and measuring health literacy in the developing world. The fundamental idea is for researchers to consider individual as well as community circumstances. These considerations are the social context of health-related decision-making, access to health information, services and resources, education and literacy, beliefs about health and health care, integration of traditional and modern health approaches, the extent to which health is a priority, as well as the speed and momentum of development. These recommendations are premised on extensive work undertaken in the low- to middle-income countries (McLaughlin and Braun 1998; Kickbusch 2001; Carrin et al. 2008; Mishra et al. 2013).

Concerning the social context of health-related decision-making, it is recommended that recognition should be given to communal actions and practices because in collectivist cultures, health decisions are made by families and not necessarily an individual. Consequently, the health literacy of household members or a household member affects other members of the family thereby impacting health outcomes. Pertaining to access to health information, services, and resources, it is contended that difference to access affects general literacy as well as health literacy. For example, individuals in places with exorbitant health costs, complex healthcare systems and hospitals, and poor availability of health service will need to draw on robust social relations to make educated decisions including behavioral and attitudinal with ramifications for health.

With regard to education and literacy, the WHO framework states that due to the likelihood of people adapting and compensating for personal inadequacies and limitations, one's level of education attained is not enough to predict and use as proxy for health literacy. The tendency to adapt enables people to develop studious ways to acquire health information by means of social relations and networks. With that in mind, any measurement tool should consider the likelihood of individuals and families developing innovative ways of engaging complex health systems given the varying capabilities that people have for engaging with health information. On beliefs about health and health care, the framework acknowledges that the cultural and religious beliefs of people impact health decision-making. With varying belief systems, what people believe due to their faith is influencing the decisions of health being made in households and by families.

In many parts of the developing world, people integrate traditional and modern health approaches in their daily healthcare practices. A thorough understanding of the processes and methods of integration is instrumental in measuring and understanding health literacy in such contexts. On the extent to which health is a priority, the perception of individual, families, communities, and central governments determines the centrality of health issues. If people perceive health as integral to other concerns of life, then in measuring health literacy, the frequency of themes emerging that are health-related may be high. That will influence the provision of opportunities to support and promote health. Finally, regarding the speed and momentum of development, consideration should be given to the pace of development on access to health and healthcare services. Several developing nations are rapidly expanding access to numerous healthcare services while introducing universal access to health. Such expansion has implications for people perceptions, knowledge, and understanding of the services being introduced and their ability to navigate them.

These theoretical discussions have guided the empirical work we have undertaken. Specifically, we have drawn on the WHO's consideration for measuring health literacy in the developing world as well as modified versions of Nutbeam's models for our studies, which are discussed subsequently in this chapter and the next.

Health Literacy on Environmental Risk to Health Among Migrants

To understand migrants' perception about environmental risks to health using survey questionnaire, we began by ascertaining their views on environmental pollution and access to pollution information in their area. Subsequently, we developed a 14-item/question instrument based on the World Health Organization's recommendation on measuring health literacy in developing countries. Specifically, the instrument measured the social context of health-related decision-making; access to health information, services, and resources; as well as education and literacy on human waste in water. Our choice of human waste as an environmental risk to health is premised on the thread of human waste as a health concern by the migrants from the participatory hazard mapping, literature that points to the presence of human excreta in the physical environment of these slums, the endemicity of cholera and other foodborne diseases in the slums, as well as our fieldwork observations. Therefore, knowledge of the dangers of human waste in water should be a concern for household decisions and actions.

Analyses

The analyses of the data consisted of descriptive statistics, computation of health literacy score, reliability analyses, validity analyses, *t*-test, ANOVA, Kruskal-Wallis tests, and Mann-Whitney tests. First, we present a description of the household responses on social context of health-related decision-making, access to health information, as well as education and literacy. Second, we continued to compute Cronbach alpha coefficients to measure the internal consistency of the scale of the entire instrument (i.e., 14 questions) used and its subscales. Afterward, a principal component analysis with varimax rotation was conducted to evaluate the factor structure of the scale. Subsequently, we examined the relationship between the health literacy variables and the household sociodemographic characteristics of the migrants after computation of health literacy scores.

Results for James Town

Descriptive Statistics

When asked whether they thought their surrounding environment (air, water, and soil) in the area where they live are polluted (dirtied with substances that are harmful to health), 64% of migrant responded in the affirmative. On access to pollution information, migrants were asked if they were aware of available information regarding pollution in the area. Forty-seven percent of migrants reported to be aware of the availability of pollution information in their area. Regarding social context of health-related decision-making, 90% of households had not discussed human waste in water 2 weeks preceding the survey, whereas 76% had not discussed human waste in water 6 months preceding the survey. Only 10% of household reported to often discuss about avoiding human waste in water during a disease outbreak in the community. With respect to health information, 68% reported that it is "not at all" easy to get leaflets on risk of drinking water containing human waste, whereas 48% indicated that it is "not at all" often to see posters/flyers on risk of drinking water containing human waste. Only 8% of respondents indicated that materials (posters/leaflets/flyers) often remind them of the dangers of drinking water containing human waste. Concerning education and literacy, 85% reported to "not at all" intentionally go out to seek (look for) information on risk of drinking water containing human waste. While 36 and 9% reported to easily interpret (easily explain the meaning of) information (drawing) on materials (posters/leaflets/flyers) about risk of drinking water containing human waste "sometimes" and "often," respectively, 31 and 8% indicated being able to easily interpret (easily explain the meaning of) information (words) on materials (posters/leaflets/flyers) about risk of drinking water containing human waste "sometimes" and "often," respectively.

Reliability Analyses

Reliability analyses were conducted to investigate the internal consistency of the scale. We achieved a very good Cronbach alpha value of 0.880 for the total scale of 14 items. We continued to evaluate the internal consistency of the subscales. To this end, the Cronbach alpha for the five items measuring "the social context of health-related decision-making" was very good (0.783), whereas the value for the five items measuring "access to health information" was also very satisfactory (0.859). The measures for the four "education and literacy" items achieved a very good value of 0.738.

Content and Construct Validity: Principal Component Analysis and Correlation

Both content and construct validity assessments were conducted. To analyze content validity about the fundamental structure of the major factors, a principal component analysis (PCA) with varimax rotation using the 14 items was undertaken. Cognizant that the items were measuring the three constituents of health literacy, this analysis was to allow us to evaluate whether each instrument item loaded onto factors as anticipated. The analysis indicated four-factor loadings that were extracted with eigenvalues greater than 1. The first, second, third, and fourth components had eigenvalues of 5.56, 1.94, 1.42, and 1.15 respectively. These four components accounted for 72% of the total variance. A closer look at the factor loading showed that while all the items measuring "the social context of health-related decision-making" and "access to health information" loaded properly on separate factors, one item measuring "education and literacy" loaded onto a fourth factor. Consequently, we dropped that item during succeeding analyses. Another PCA using the remaining 13 items was conducted. At this stage, the analyses showed three factor loadings that were extracted with eigenvalues greater than 1 (see Table 5.1). The eigenvalues of the components are 5.40, 1.78, and 1.28 for the first, second, and third components, respectively, and these accounted for 65% of the total variance. After the rotation, the "education and literacy" items all loaded onto factor 1 and "the social context of health-related decision-making" all loaded onto the second factor. The remaining items on "access to health information" loaded onto factor 3. Concerning construct validity, the relationship of subscales scores to each other was assessed. Scores on "the social context of health-related decision-making" items were significantly correlated with "access to health information" items ($r = 0.445$, $p < 0.001$) and significantly associated with "education and literacy" items ($r = 0.435$, $p < 0.001$). A statistically significant association between "access to health information" and "education and literacy" items was also found ($r = 0.623$, $p < 0.001$).

Bivariate Analysis: Health Literacy Scores and Demographic Factors

There were significant associations between health literacy scores and employment status, educational attainment, religiosity, marital status, and occupation of respondents (Table 5.2). Employment status was associated with "access to information" with the unemployed having a higher mean score than their employed counterparts. Education level was statistically associated with "total health literacy score," "access to information," and "education and literacy." In all cases, higher education was related with higher health literacy scores. Those with no education consistently scored lower than those with primary, secondary, or higher education. Religiosity was significantly associated with "total health literacy score," "the social context of health-related decision-making," "access to health information," and "education and literacy." While Mann-Whitney test did not reveal a significant difference between respondents who reported religiosity as "not at all" and those who reported theirs as always, there was significant relationship between all other categories. For example, those who reported to always practice the tenets of their religious beliefs had higher mean rank scores on total health literacy and the subscales than those who occasionally do. Married and single migrants did not have statistically significant difference in their score on health literacy. However, married respondents had higher scores on "total health literacy score" and "access to health information" compared to migrants who were divorced or separated. Migrants in professional/managerial occupations

Table 5.1 Instrument item scores and principal component analysis factor loadings (James Town)

Instrument items	Means (SD)	Principal component analysis Factor loadings 1	2	3
How often in the past 2 weeks have you talked about human waste in water in your household?	1.17 (0.552)	−0.107	0.519	0.512
How often do you or anyone in your household adhere to suggestions from other household members on how to prevent drinking water containing human waste?	1.62 (0.868)	0.124	0.547	0.329
How often in the past 6 months have you discussed human waste in water as a household?	1.4 (0.763)	0.109	0.72	0.428
How often do you decide as a household on steps to take to avoid drinking water containing human waste?	1.8 (1.092)	0.291	0.758	−0.061
How often does your household talk about avoiding human waste in water during a disease outbreak in this community?	1.88 (1.081)	0.236	0.86	−0.01
How easy is it for you (or anyone in your household) to get/receive leaflets on risk of drinking water containing human waste?	1.51 (0.829)	0.323	0.01	0.779
How often do you (or anyone in your household) see posters/flyers on risk of drinking water containing human waste?	1.8 (0.880)	0.416	0.07	0.731
How easy is it for you (or anyone in your household) to understand the leaflets on risk of drinking water containing human waste?	2.16 (1.095)	0.791	0.134	0.356
How often do these materials (posters/leaflets/flyers) remind you (or anyone in your household) of the dangers of drinking water containing human waste?	2.12 (1.016)	0.696	0.124	0.381
Are you (or anyone in your household) the sort of person who uses the information on the materials (posters/leaflets/flyers) to teach other household members risk of drinking water containing human waste?	1.91 (1.032)	0.731	0.159	0.224
How often do you (or anyone in your household) intentionally go out to seek (look for) information on risk of drinking water containing human waste?	1.24 (0.610)	0.11	0.185	0.494
Are you the sort of person (or anyone in your household) who can easily interpret (easily explain the meaning of) information (drawing) on materials (posters/leaflets/flyers) about risk of drinking water containing human waste?	2.16 (1.048)	0.857	0.259	0.069
Are you the sort of person (or anyone in your household) who can easily interpret (easily explain the meaning of) information (words) on materials (posters/leaflets/flyers) about risk of drinking water containing human waste?	2.07 (1.015)	0.88	0.128	0.058

scored significantly higher on "total health literacy score," "the social context of health-related decision-making," "access to health information," and "education and literacy" compared to those in sales/services and manual jobs.

Table 5.2 Bivariate analysis of health literacy scores and demographic factors (James Town)

	Total health literacy score	Social context of health decision-making	Access to health information	Education and literacy
Employment status	1.706	0.931	2.082*	0.938
Education level	4.909*	1.011	5.822**	5.205***
Religiosity	6.775**	3.326***	9.049**	2.739*
Marital status	3.105*	1.183	2.778*	2.662*
Occupation	4.931***	3.596*	3.847*	2.695*

*$p < 0.05$, **$p < 0.001$, ***$p < 0.01$

Results for Old Fadama

Descriptive Statistics

Seventy-nine percent of migrant respondents asserted that their surrounding environment (air, water, and soil) in the area where they live are polluted. On access to pollution information, 51% of migrants reported to be aware of the availability of pollution information in their area. With respect to social context of health-related decision-making, 88% of households had not discussed human waste in water 2 weeks preceding the survey, whereas 67% had not discussed human waste in water 6 months preceding the survey. Thirteen percent of household reported to often discuss about avoiding human waste in water during a disease outbreak in the community. Concerning access to health information, 64% reported that it is "not at all" easy to get leaflets on risk of drinking water containing human waste, whereas 53% indicated that it is "not at all" often to see posters/flyers on risk of drinking water containing human waste. Only 5% of respondents indicated that materials (posters/leaflets/flyers) often remind them of the dangers of drinking water containing human waste. Relating to education and literacy, 83% reported to "not at all" intentionally go out to seek (look for) information on risk of drinking water containing human waste. While 20 and 6% reported to easily interpret (easily explain the meaning of) information (drawing) on materials (posters/leaflets/flyers) about risk of drinking water containing human waste "sometimes" and "often," respectively, 17 and 7% indicated being able to easily interpret (easily explain the meaning of) information (words) on materials (posters/leaflets/flyers) about risk of drinking water containing human waste "sometimes" and "often," respectively.

Reliability Analysis

To investigate the internal consistency of the scale, reliability analyses were conducted. For the total scale of 14 items, a very good Cronbach alpha value of 0.832 was achieved. The internal consistency of the subscales was also examined. To this end, the Cronbach alpha for the five items measuring "the social context of health-related decision-making" was very good (0.898), whereas the value for the five items measuring "access to health information" was also very good (0.817). The measures for the four "education and literacy" items achieved an adequate value of 0.687.

Content and Construct Validity: Principal Component Analysis and Correlation

Content and construct validity assessments were conducted. A principal component analysis (PCA) with varimax rotation using the 14 items was undertaken to examine content validity about the fundamental structure of the major factors. Knowing that the items were measuring the three major elements of health literacy, this analysis was to allow us to evaluate whether each instrument item loaded onto factors as expected. The analysis indicated a four-factor loading that were extracted with eigenvalues greater than 1. The first, second, third, and fourth components had eigenvalues of 4.54, 3.16, 1.51, and 1.15, respectively. These four components accounted for 74% of the total variance. The factor loadings reveal that while all the items measuring "the social context of health-related decision-making" and "access to health information" loaded properly on separate factors, two items measuring "education and literacy" loaded onto a fourth factor. Consequently, we dropped those items during subsequent analyses. Another PCA using the remaining 12 items was conducted. At this stage, the analyses showed three factor loadings that were extracted with eigenvalues greater than 1 (Table 5.3). The eigenvalues of the components are 4.37, 3.00, and 1.19 for the first, second, and third components, respectively, and these accounted for 71% of the total variance. After the rotation, the "social context of health-related decision-making" items all loaded onto factor 1 and the "access to health information" all loaded onto the second factor. The remaining items on "education and literacy" loaded onto factor 3. Regarding construct validity, the relationship of subscale scores to each other was assessed. Although scores on "the social context of health-related decision-making" items were not significantly correlated with "access to health information" items as well as the "education and literacy" items, a statistically significant association between "access to health information" and "education and literacy" items was found ($r = 0.568$, $p < 0.001$).

Bivariate Analysis: Health Literacy Scores and Demographic Factors

The demographic variables that were significantly associated with health literacy in Old Fadama include sex, duration of stay, household size, and educational attainment (Table 5.4). Sex was significantly associated with "social context of health-related

Table 5.3 Instrument item scores and principal component analysis factor loadings (Old Fadama)

Instrument items	Means (SD)	Principal component analysis Factor loadings 1	2	3
How often in the past 2 weeks have you talked about human waste in water in your household?	1.14 (0.402)	0.586	0.218	0.184
How often do you or anyone in your household adhere to suggestions from other household members on how to prevent drinking water containing human waste?	1.59 (0.906)	0.881	0.032	0.065
How often in the past 6 months have you discussed human waste in water as a household?	1.54 (0.872)	0.902	−0.026	0.12
How often do you decide as a household on steps to take to avoid drinking water containing human waste?	1.71 (0.936)	0.931	0.015	0
How often does your household talk about avoiding human waste in water during a disease outbreak in this community?	1.97 (1.108)	0.894	0.067	−0.025
How easy is it for you (or anyone in your household) to get/receive leaflets on risk of drinking water containing human waste?	1.43 (0.628)	0.092	0.744	0.062
How often do you (or anyone in your household) see posters/flyers on risk of drinking water containing human waste?	1.57 (0.679)	−0.005	0.855	−0.006
How easy is it for you (or anyone in your household) to understand the leaflets on risk of drinking water containing human waste?	1.77 (0.901)	0.121	0.606	0.536
How often do these materials (posters/leaflets/flyers) remind you (or anyone in your household) of the dangers of drinking water containing human waste?	1.76 (0.925)	0.03	0.734	0.336
Are you (or anyone in your household) the sort of person who uses the information on the materials (posters/leaflets/flyers) to teach other household members risk of drinking water containing human waste?	1.62 (0.887)	0.084	0.618	0.353
Are you the sort of person (or anyone in your household) who can easily interpret (easily explain the meaning of) information (drawing) on materials (posters/leaflets/flyers) about risk of drinking water containing human waste?	1.91 (0.922)	0.215	0.186	0.885
Are you the sort of person (or anyone in your household) who can easily interpret (easily explain the meaning of) information (words) on materials (posters/leaflets/flyers) about risk of drinking water containing human waste?	1.75 (0.962)	−0.027	0.21	0.913

decision-making" with females having higher score compared to their male counterparts. The length of stay was inversely correlated with scores on "access to health information," whereas household size was positively correlated with "total health literacy score" and "social context of health-related decision-making." Migrant respondents with no education consistently scored lower than those with primary, secondary, or higher education. Those with secondary or higher education had significantly higher

Table 5.4 Bivariate analysis of health literacy scores and demographic factors (Old Fadama)

	Total health literacy score	Social context of health decision-making	Access to health information	Education and literacy
Sex	0.504	2.046*	−0.601	−0.751
Duration of stay	−0.156	−0.017	−0.198*	−0.153
Household size	0.189*	0.170*	0.141	0.07
Education level	11.961***	2.639	16.374***	16.980***

*$p < 0.05$, **$p < 0.01$, ***$p < 0.001$

scores in "total health literacy," "social context of health-related decision-making," "access to health information," as well as "education and literacy" than migrants with primary and secondary school education.

Brief Deliberation on Health Literacy on Pollution and Human Waste in Water

Most residents in both James Town and Old Fadama agree that their environment is polluted. However, as expected based on the deplorable exogenous factors in Old Fadama illustrated with pictures in Chap. 1, higher percentage of the residents in Old Fadama affirmed their polluted surroundings compared to their counterparts in James Town. That notwithstanding, low proportions of migrant households discuss the dangers of human waste (excreta and urine) in water in both slums even in the event of major diarrheal disease outbreak in the communities.

On access to health information, over half of migrant respondents thought it was not easy to access leaflets on dangers of human waste in water or see posters on risks of drinking water contaminated with human waste. This coupled with the low levels of intentionality regarding seeking health information, we opine, is not only indicative of poor health literacy but also potential conduit for negative health outcomes. This may be partially responsible for the high levels of typhoid fever recorded in Old Fadama as discussed in Chap. 3.

General literacy positively impacted health literacy. Migrants with higher education had better scores on all typologies of health literacy in both slums. That is, general literacy in these contexts is facilitating household health-related decision-making and access to health information, which may exert positive influence on household members thereby positively affecting health outcome. Socioeconomic status significantly affected health literacy. Migrants in professional or clerical occupations, which are proxies for high socioeconomic status, scored better on health literacy compared to other migrants in manual work/occupations. In the next section, we delve deeper into perception of fecal contamination in the slums, which will provide more information about the low level of health literacy on human waste discussed.

Conceptual Mapping (Mental Models) on Human Waste (Fecal Contamination)

In this section, we present the results from the conceptual mapping on fecal contamination in all the four communities in Ghana and Uganda. This activity enabled us to understand how participants and medical/health personnel think about or understand the causes/drivers and consequences of fecal contamination. By way of the step-by-step process of this activity, firstly, fecal contamination was written in the middle of a large sheet of paper. Participants were asked to discuss and brainstorm the reasons for and causes of fecal contamination in their community as well as the source of their information. Each response of a cause or reason was listed on sticky notes and attached to the "cause" side of the large paper. The discussion surrounding the reasons/causes listed resulted in connecting arrows from the reasons to the problem (fecal contamination) and to each other based on respondents understanding of the connections among the causes and between the causes and the problem. Subsequently, participants were asked to discuss and brainstorm the consequences of fecal contamination for people and the environment. The responses were listed on sticky notes and attached to the consequences side of the large paper. The perspectives on the consequences mentioned led the connecting arrows from the problem (fecal contamination) to consequences on people and the environment.

Results from Mental Models in James Town

In James Town, the main reasons and causes of fecal contamination in the community identified by residents and health professionals are inadequate (lack of) toilet facilities and absence of nonshared toilet facilities in households. While resident migrants mentioned other drivers of fecal contamination such as inability to afford the regular use of the public toilet, children refusal to use public toilets, the closure of public toilets during certain times of the day, and no space on the landscape to put up toilet facility in the household by landlords, the health professional provided further nuanced linkages and noted that perceived lack of sanitation education is primarily responsible for indiscriminate disposal of feces in the community. The problem of the closure of the public toilets was explained by one of the participants as follows:

> Another issue too has to do with the locking up of the toilet facility at night. At 11:00pm, the one in charge locks it up and goes home to avoid persons from using it since he will not be available to collect the money [payment for using the facility]. It is closed between the hours of 11:00pm in the night to 04:00am at dawn. The workers cannot sit there always so they prefer to keep it under lock when unavailable. [Female; 40 years; 15 years in community]

The responses of migrants and health professional on the consequences of fecal contamination converged on two principal issues: (1) sickness—for consequences on people, and (2) unsanitary conditions—for consequences on the environment. Migrant residents mainly indicated that the consequence of fecal contamination for people is sickness such as diarrhea, cholera, and malaria. The health professionals buttressed these consequences, except malaria, and added more multipart aspects including other waterborne diseases, typhoid, and worm infestation. Regarding the unsanitary conditions as a consequence of fecal contamination on the environment, while resident migrants mainly mentioned filthy surroundings and microscopic organism such as bacteria, the health professionals explained further with implications for hygiene and pollution. See Figs. 5.1 and 5.2 for the aggregate conceptual model of fecal contamination for James Town and pictures from some of the migrant and health professional respondents, respectively.

Results from Mental Models in Old Fadama

The main causes of fecal contamination in Old Fadama recognized by both migrants and health professionals are inadequate (lack of) toilet facilities and the unhygienic toilet facilities in the area. Other reasons migrants mentioned as drivers of fecal contamination are low levels of educational attainment by residents, the relative high cost associated with the use of the toilet facilities in the area, the limited drainage system, rampant littering and dumping habits, early closure of toilet facilities, inadequate refuse dumps, no government-sanctioned toilet facilities, and sheer wickedness on the part of some residents (Fig. 5.3). The health professional added that the overcrowding in the community puts extra strain on the already limited and unhygienic facilities leading to spill overs and, consequently, fecal contamination in the neighborhood. While most of the reasons given in Old Fadama mirror that of James Town, we see a few differences. Regarding similarities, one that is quite captivating is the early closure of the facilities. A participant explained that phenomenon and its attendant consequence as follows:

> The owners of the toilet facilities usually close early and this makes people [individuals with no toilet facilities in their household or near them] defecate anywhere and also throw their faces anywhere after they have eased themselves. [Male; 43 years; 8 years in community]

The inadequate toilet facilities and associated issues allows residents to devise ways to dispose off human waste in ways that increase the susceptibility of residents to hazards related to improper disposal of night soil. A migrant lamented such practices:

> [There is] no refuse dump for the community to dispose of refuse… after they have made their waste, which therefore makes them dispose refuse like feces, which they wrapped in plastic bags, around or anyplace they find available in the community. [Male; 27 years; 6 years in community]

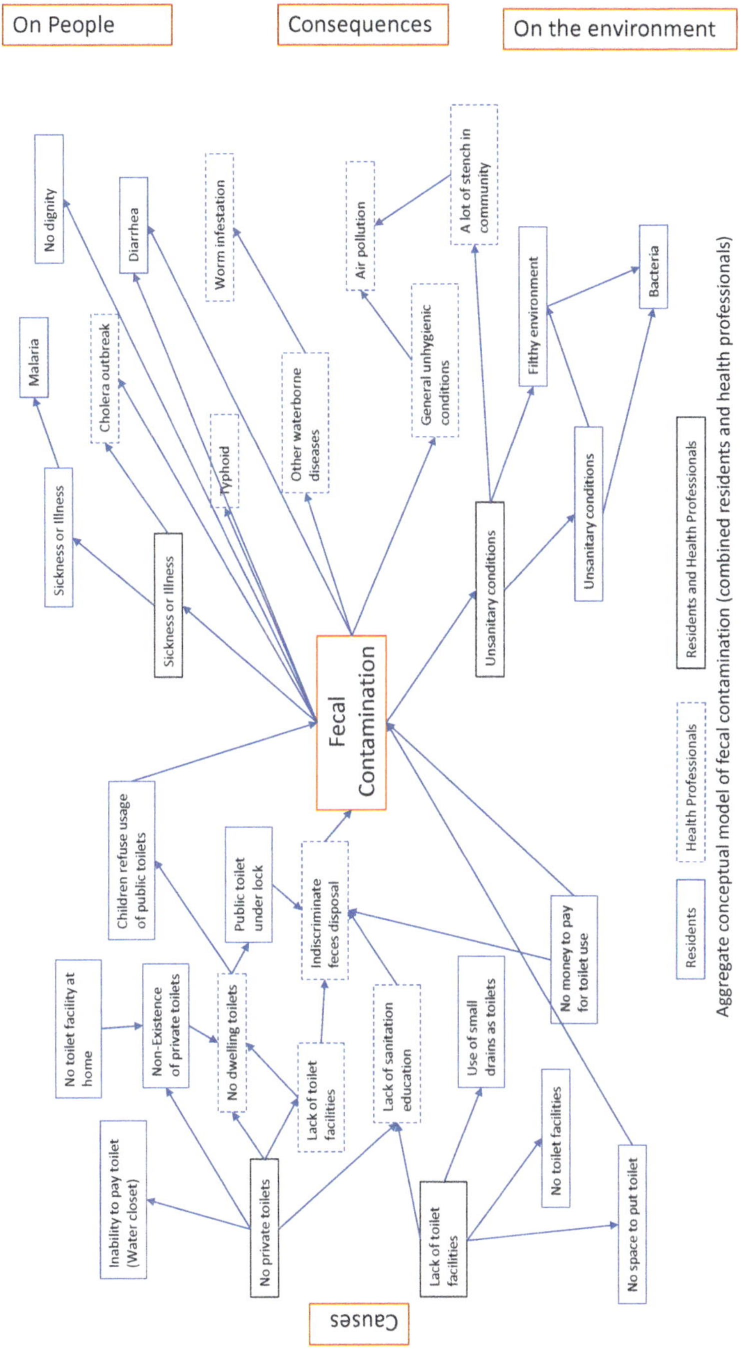

Fig. 5.1 Conceptual Model of Fecal Contamination, James Town

Fig. 5.2 Conceptual models on fecal contamination, James Town

Fig. 5.3 Conceptual models on fecal contamination, Old Fadama

For those who may not defecate in their housing structures and dispose of it clandestinely, they chose to go out in the open and defecate at will. One of the health professionals remarked:

> Community members usually engage in behaviors that result in fecal contamination because of the unhygienic nature of the available public toilet facilities; [the unhygienic nature] usually drives them away from using them [Male; 23 years; 1 year in community]

The habit of open defection and indiscriminate dumping of feces by residents were sometimes regarded as intentional misbehavior and or intentional acts of indiscretion. A migrant irritably put it this way:

> It is the uncivilized behavior on the part of most community members. Most community members are cruel, and this situation usually makes them engage in such acts. [Male; 43 years; 8 years in community]

Like James Town, the responses of migrants and health professional on the consequences of fecal contamination yielded two main themes: (1) diseases—for effects on people, and (2) unhygienic conditions—for effects on the environment. The consequences of fecal contamination on people mentioned by migrant participants included diarrhea, vomiting, stomach aches and upset, malaria, cholera, loss of appetite, and candida.

We find the loss of appetite quite intriguing. This was explained as:

> Community members and even outsiders can lose their appetite when they see such things in the community. [Male; 43; 8 years in community]

The health professionals' responses supported most of these diseases and further displayed the pathways by which improper night soil disposal connects with disease transmission and increased risk for infections. Migrants and health professionals agreed on unhygienic impacts of fecal contamination. While their immediate environment is not very pleasant, the surroundings are further polluted by feces and in the event of flooding, fecal contaminants block the free flow of water in their scanty drains and further seep into housing structures. A health professional explained this as follows:

> The whole community can be flooded. This is because, the feces especially the ones in the plastic bags that have been dispose into the gutters can make the gutters choked which prevents the flow of water whenever it rains. [Male; 23 years; 1 year in community]

Results from Mental Models, Kinawataka

In Kinawataka, the causes of fecal contamination identified by residents and health professionals are poor waste disposal and poor sanitation in the area (Figs. 5.4, 5.5, and 5.6). Health professionals also indicated that during the rainy seasons, the water channel fills up and floods, leading to contamination of all most water sources in the community. Some participants said it was due to negligence because some residents defecate in polythene bags and dispose off in public open places or in garbage bins leading to contamination of water sources and the environment. Health professionals also argued that the fecal contamination is a direct result of the poor standards of living as explained below:

> People in katoogo are very poor. The houses leak and they cannot even afford to visit a public toilet. Due to poverty and congestion, they have no space to construct toilets because most of them are renting the houses where they are staying. They end up using polythene bags and buckets as toilets leading to contamination of the water sources and the environment. [Female; 50 years; 7 years in the Community]

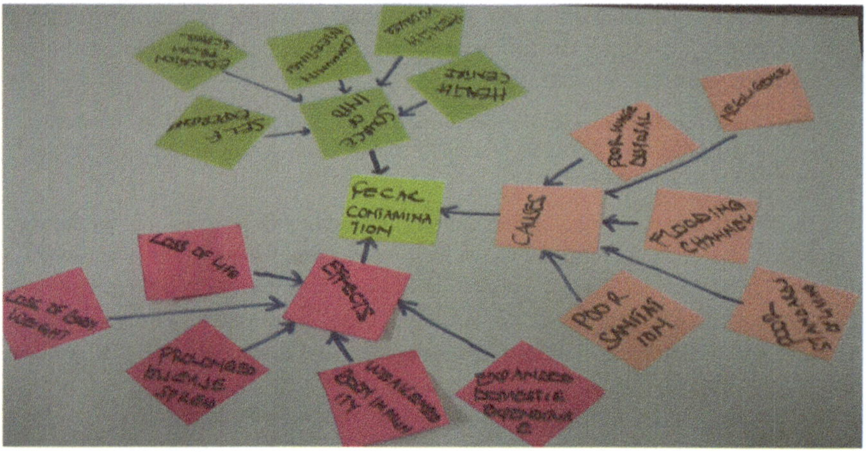

Fig. 5.4 Conceptual models on fecal contamination, Kinawataka

Fig. 5.5 Conceptual mapping on fecal contamination, Kinawataka

Residents and health professionals also indicated that fecal contamination often leads to high expenses per household overtime due to the recurrence of diseases, many times leading to loss of lives. Diseases singled out as the most common are diarrhea related—cholera and dysentery; and typhoid. Health professionals explained that sometimes, especially, children get such diseases continuously putting a strain on households to the extent that some are unable to afford the treatment. She noted that:

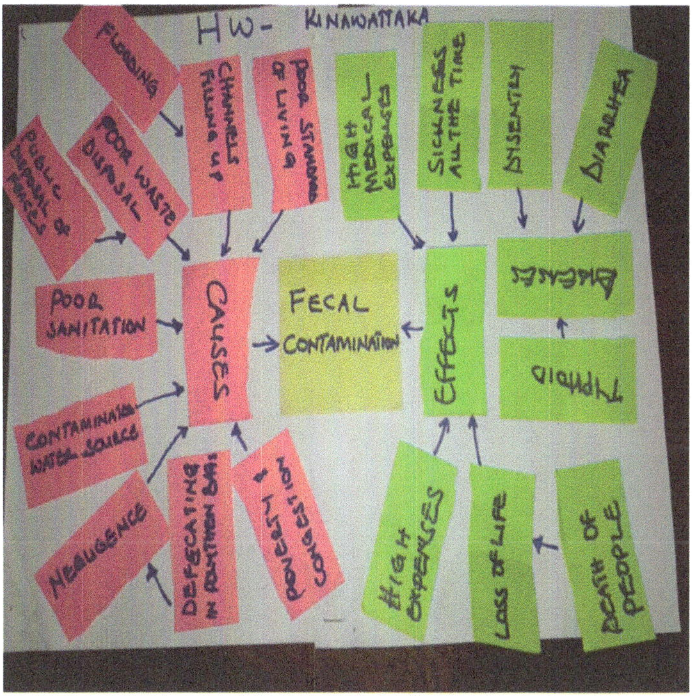

Fig. 5.6 Conceptual mapping on fecal contamination (health workers) in Kinawataka

> Sometimes the same children are brought to the clinic almost everyday for two weeks due to diarrhoeal diseases leading to dehydration and sometimes death. Due to lack of money sometimes it is not possible to pay for treatment all the time leading to deaths of children. [Female health worker; 40 years]

According to the participants, it was needs such as health care that did not leave any resources for buying anything else or affording a better house, hence poor standard of living in the area. In other words, the cost of health care seems to eating a huge chunk of their earnings hence their inability to save toward purchasing items that will provide them with material comfort.

Results from Mental Models, Namuwongo

Fecal contamination in Namuwongo was mainly attributed to ungazetted dumping grounds, poor waste disposal, lack of enough toilet facilities, and fetching water from open sources (Fig. 5.7). Residents reported that people tend to use "flying toilets" and the bucket system especially those who have no access to toilets at night. The health professionals added that pit latrines are so dirty and sometimes maggots can be seen with naked eyes on top of the latrines and yet the houses are so congested.

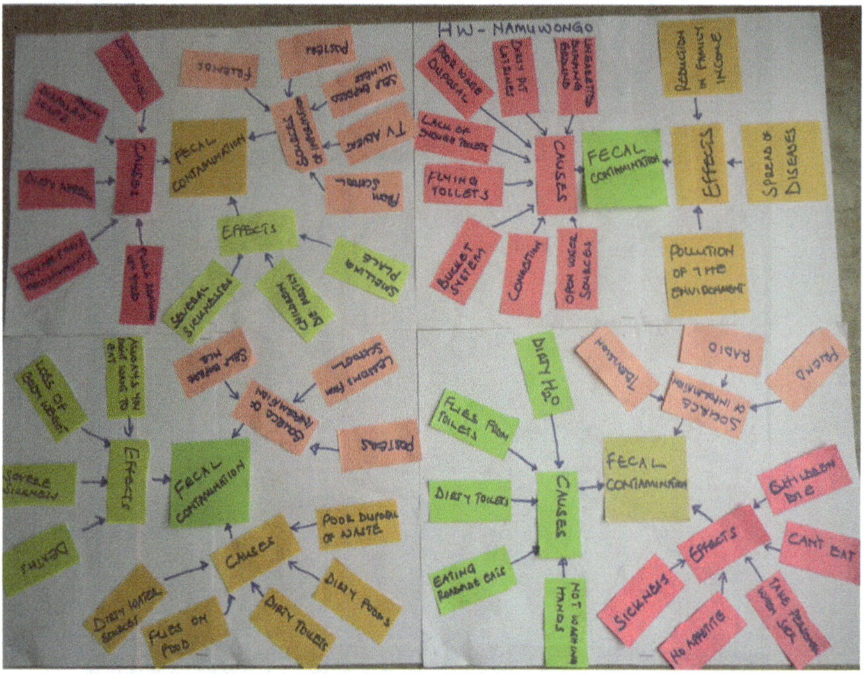

Fig. 5.7 Conceptual model on fecal contamination, Namuwongo

Residents reported that fecal contamination often leads to pollution of the environment, spread of diseases, and reduction in the incomes of the affected families.

In both slums, reasons given were similar in nature, especially about toilet usage and lack of toilets and the consequential fecal matter disposal habits as explained by one participant:

> People do not have toilet facilities in their houses, so at night when one needs to use a toilet, they simply do everything in a bucket or polythene bag and the next day wake up early in the morning to throw it anywhere in the neighborhood as long as no one is seeing them. [Male; 46 years; 10 years in community]

For those who have toilet facilities, during the rainy season:

> They open the outlets of the toilets whenever there is a flood because there is no other way of emptying such pit latrines that are not planned and are not even accessible by anyone that would be emptying them. Sometimes, when our houses flood you see feces also floating in the water [Female; 35 years; 6 years in the Community]

Most participants insinuated in so many words that it is a choice between keeping feces smelling in their houses and finding a clever way to dispose it off without being seen. Most of the diseases that affect the community were attributed to the improper disposal of fecal matter in the community by the health workers.

Brief Deliberation on Mental Models on Fecal Contamination

The conceptual mapping activities sought to understand migrants' perceptions on the drivers of fecal contamination in the slums and their associated consequences on people and the environment. Mirroring most very poor and deprived slums across the globe, the primary causes of fecal contaminant, according to the migrants, are lack of or inadequate toilet facilities, nonexistent or inadequate nonshared toilet facilities, unavailability of toilet facilities due to closure during the day, fetching water from open sources, and insufficient space to build toilets. The discussions on the consequences of fecal contamination on people and the environment suggested two main effects—diseases and unsanitary conditions.

Although all the slums seem to have similar problems, the difficulties play out differently across space and differ in severity. For example, while participants across the slums mentioned inadequate toilet facilities, it is more severe in the squatter settlement. Old Fadama, for instance, seem to have the unguided and nonexistent government-supervised toilet because the slum is not a designated residential area, hence experience frequent fecal spill over from the private-owned toilet facilities. Coupled with the poor drainage system, the presence of fecal matter around is frequent and palpable.

Perceptually and behaviorally, the discussion revealed the idea of residents treating feces like any other refuse may be a major cause for concern. It is understood that one may prefer to dispose off feces to avoid smell in the home. However, the idea of flying it across the streets or in choked drains show a lack of deeper understanding of the consequences of fecal contamination by many residents. This may confirm the assertion by some of the health professionals that such behavior is attributable to lack of education on the proper disposal of feces and connections to various diseases. More so, although mention is made of diseases such as cholera, dysentery, and typhoid, other predominant diseases associated with fecal contamination are left out. These include hepatitis, ascariasis, and gastroenteritis.

References

Baker DW (2006) The meaning and the measure of health literacy. J Gen Intern Med 21(8):878–883

Bann CM, McCormack LA, Berkman ND, Squiers LB (2012) The health literacy skills instrument: a 10-item short form. J Health Commun 17(suppl 3):191–202. https://doi.org/10.1080/10810730.2012.718042

Bennett IM, Chen J, Soroui JS, White S (2009) The contribution of health literacy to disparities in self-rated health status and preventive health behaviors in older adults. Ann Fam Med 7(3):204–211. https://doi.org/10.1370/afm.940

Carrin G, Mathauer I, Xu K, Evans DB (2008) Universal coverage of health services: tailoring its implementation. Bull World Health Organ 86(11):857–863

Chinn D, McCarthy C (2013) All Aspects of Health Literacy Scale (AAHLS): developing a tool to measure functional, communicative and critical health literacy in primary healthcare settings. Patient Educ Couns 90(2):247–253. https://doi.org/10.1016/j.pec.2012.10.019

Chung SY, Nahm ES (2015) Testing reliability and validity of the eHealth Literacy Scale (eHEALS) for older adults recruited online. Comput Inform Nurs 33(4):150–156. https://doi.org/10.1097/cin.0000000000000146
DeWalt DA, Berkman ND, Sheridan S, Lohr KN, Pignone MP (2004) Literacy and health outcomes. a systematic review of the literature. J Gen Intern Med 19:1228–1239
Educational Testing Service (2006) Test content for health activities literacy tests. https://www.ets.org/. Accessed 9 May 2018
Institute of Medicine (2004) Health literacy: a prescription to end confusion. National Academies Press, Washington, DC
Kickbusch IS (2001) Health literacy: addressing the health and education divide. Health Promot Int 16(3):289–297
Kim MT, Song HJ, Han HR, Song Y, Nam S, Nguyen TH et al (2012) Development and validation of the high blood pressure-focused health literacy scale. Patient Educ Couns 87(2):165–170. https://doi.org/10.1016/j.pec.2011.09.005
Lee SY, Arozullah AM, Cho YI (2004) Health literacy, social support, and health: a research agenda. Soc Sci Med 58(7):1309–1321. https://doi.org/10.1016/S0277-9536(03)00329-0
Liu YB, Liu L, Li YF, Chen YL (2015) Relationship between health literacy, health-related behaviors and health status: a survey of elderly Chinese. Int J Env Res Public Health 12(8):9714–9725
McLaughlin LA, Braun KL (1998) Asian and Pacific Islander cultural values: considerations for health care decision making. Health Soc Work 23(2):116–126
Mishra S, Kusuma YS, Babu BV (2013) Concepts of health and illness: continuity and change among migrant tribal community in an Eastern Indian city. Anthropol Notebooks 19(3):61–69
Murphy PW, Davis TC, Long SW, Jackson RH, Decker BC (1993) Rapid estimate of adult literacy in medicine (REALM): a quick reading test for patients. J Read 37(2):124–130
Nutbeam D (2000) Health literacy as a public health goal: a challenge for contemporary health education and communication strategies into the 21st century. Health Promot Int 15(3):259–267
Nutbeam D (2008) The evolving concept of health literacy. Soc Sci Med 67(12):2072–2078
Paasche-Orlow MK, Wolf MS (2007) The causal pathways linking health literacy to health outcomes. Am J Health Behav 31(1):S19–S26
Parker RM, Baker DW, Williams MY, Nurss JR (1995) The Test of Functional Health Literacy in Adults (TOFHLA): a new instrument for measuring patient's literacy skills. J Gen Intern Med 10(10):537–541. https://doi.org/10.1007/bf02640361
Ratzan SC, Parker R, Selden CR, Zorn M (2000) National Library of medicine current bibliographies in medicine: health literacy. National Institutes of Health, Bethesda. https://www.researchgate.net/publication/230877250_National_Library_of_Medicine_Current_Bibliographies_in_Medicine_Health_Literacy
Schillinger D, Grumbach K, Piette J, Wang F, Osmond D, Palacios J et al (2002) Association of health literacy with diabetes outcomes. J Am Med Assoc 288:475
Sørensen K, Van den Broucke S, Fullam J, Doyle G, Pelikan J, Slonska Z, Brand H (2012) Health literacy and public health: a systematic review and integration of definitions and models. BMC Public Health 12(1):80
Speros C (2005) Health literacy: concept analysis. J Adv Nurs 50(6):633–640
Suka M, Odajima T, Okamoto M, Sumitani M, Igarashi A, Ishikawa H, Kusama M, Yamamoto M, Nakayama T, Sugimori H (2015) Relationship between health literacy, health information access, health behavior, and health status in Japanese people. Patient Educ Couns 98(5):660–668
Weiss BD, Mays MZ, Martz W, Castro KM, DeWalt DA, Pignone MP et al (2005) Quick assessment of literacy in primary care: the newest vital sign. Ann Fam Med 3(6):514–522. https://doi.org/10.1370/afm.405
Wolf MS, Williams MV, Parker RM, Parikh NS, Nowlan AW, Baker DW (2007) Patients' shame and attitudes toward discussing the results of literacy screening. J Health Commun 12(8):721–732. https://doi.org/10.1080/10810730701672173
World Health Organization (2015) Health literacy toolkit for low- and middle-income countries: a series of information sheets to empower communities and strengthen health systems. WHO, Geneva

Chapter 6
Health Literacy of Migrants: Infectious and Noncommunicable Diseases

Abstract This chapter examines migrant's health literacy on communicable and noncommunicable diseases. Based on the most reported infectious disease, apart from malaria, and noncommunicable disease from the immediate clinics in Old Fadama and James Town, a survey questionnaire was developed to measure migrant's functional, communicative/interactive, and critical health literacy based on Nutbeam's health literacy typologies. The questionnaire was administered to a random sample of 150 individuals per site. The data were analyzed using descriptive statistics, reliability analyses, validity analyses, independent samples *t*-test, Pearson correlation, and analysis of variance (ANOVA). The instruments were reliable and achieved very good Cronbach Alpha values. The results indicate that, in James Town, there is a low level of functional, communicative, and critical health literacy among the study respondents regarding infectious disease. Similar trend is found in Old Fadama although migrants in Old Fadama scored better on communicative health literacy regarding upper respiratory disease. Hypertension was the noncommunicable disease mostly reported in both slums. Health literacy was found to be demographically differentiated. Generally, the higher one's general literacy, the better their scores on health literacy.

Keywords Health literacy · Infectious disease · Noncommunicable disease · Old Fadama · James Town · Hypertension · Respiratory diseases · Functional literacy · Communicative literacy · Critical literacy

Introduction

Paasche-Orlow and Wolf (2007) postulated that the mediating pathways between health literacy and health outcomes is a function of both individual and healthcare system experiences. Consequently, individual background factors such as ethnicity,

R. A. Tutu, J. D. Busingye, *Migration, Social Capital, and Health*, Global Perspectives on Health Geography,
https://doi.org/10.1007/978-3-030-24693-8_6

education, age, occupation, income, employment, social capital, and culture influence health literacy with implications for health outcomes. So, to assess the extent to which an individual can obtain, process, understand, and talk about health information, consideration should be given various facets of health literacy (e.g., print, oral literacy, numeracy, information-seeking behavior) and contextualized (Bann et al. 2012; Chinn and McCarthy 2013; Chung and Nahm 2015; Kim et al. 2012). Migrants' health literacy on common infectious and noncommunicable diseases in the communities are examined in this chapter. We draw on Chinn and McCarthy (2013), who having identified drawbacks in the existing health literacy measurement tools utilized Nutbeam (2000) conceptualization of health literacy to develop a self-reported tool that focused on understanding health literacy in social and institutional context. It is worthy of note that all migrants interviewed had heard or experienced the infectious or noncommunicable disease in their respective communities. Following the data from the clinics serving the communities and available health literature, the body health mapping, and participatory hazard mapping activities, as discussed previously in Chaps. 3, 4, and 5, a specific infectious disease was chosen to explore in each community. For James Town, we chose a sexually transmitted diseases, gonorrhea, while in Old Fadama, an upper respiratory infection (common cold) was chosen to examine migrants' health literacy.

Health Literacy on Infectious Diseases Among Migrants

We developed a nine self-report items/questions instrument based Nutbeam's (2000) model, which discusses three health literacy typologies: functional, communicative/interactive, and critical health literacy. Functional health literacy denotes the person's basic capability to read and write including control over basic knowledge of health circumstances. Communicative/interactive health literacy represents social skills that are employed in the acquisition of health information and ability to construe meaning from diverse communication forms and its application to numerous variable situations. Critical health literacy signifies an advanced level of social and reasoning abilities that allows the person to analytically appraise health information and by so doing enables him/her to exercise control over their life course. Each of the nine items had the following response options and codes: "not at all,"—1; "rarely,"—2; "sometimes,"—3; and "often"—4.

Analyses

The analyses of these data comprised of descriptive statistics, reliability analyses, validity analyses, independent samples *t*-test, Pearson correlation, and analysis of variance (ANOVA). A description of the percentage distribution of the responses to the nine items measuring functional health literacy, interactive/communicative health literacy, and critical health literacy is discussed. An assessment of the internal consistency of the scale of the whole instrument as well as the subscales was undertaken through a computation of Cronbach alpha coefficients. Next, a principal component analysis using varimax rotation was undertaken to evaluate the factor structure of the scale. We computed health literacy scores and subsequently assessed the relationship between health literacy and demographic characteristics using ANOVA, Pearson correlation, and independent samples *t*-test.

Results on Health Literacy on Gonorrhea in James Town

Descriptive Statistics

With regard to functional health literacy items, while about 65% of migrant respondents indicated that they do not need help when they are given information on gonorrhea to read by a doctor, nurse or pharmacist, 22% and 7% reported to require help "sometimes" and "often," respectively. Twenty-six and 9% reported that they can easily get help to comprehend information on gonorrhea "sometimes" and "often," respectively. On communicative health literacy, while about 67% indicated that they do not always provide the nurse or doctor all the information they need when they talk to them about gonorrhea, 13 and 12% reported to "sometimes" and "often" ask all the questions they need to ask the doctor or nurse, respectively, when they talk to them. Regarding critical health literacy, when asked whether they like to find out lots of different information about your health regarding gonorrhea, 73% indicated "not at" while 13 and 6% indicated "sometimes" and "often," respectively. When asked about how often they thought carefully about whether health information on gonorrhea made sense in a particular situation, 69% indicated "not at all." Meanwhile, 16 and 4% reported "sometimes" and "often," respectively.

Reliability Analyses

Reliability analyses were performed to investigate the internal consistency of the nine-item scale. A very good Cronbach alpha value of 0.844 was attained for the total scale of nine items. With respect to each of the subscales, the Cronbach alpha value for the items assessing functional health literacy was adequate (0.646), while the Cronbach alpha value for the three items evaluating communicative health literacy was excellent (0.952). The analyses on the measure for the critical health literacy items attained a very good value of 0.876.

Content and Construct Validity: Principal Component Analysis and Correlations

To investigate content validity with respect to the fundamental structure of the nine factors, a principal component analysis with varimax rotation was performed. Since the items were assessing the three health literacy typologies, this analysis was to facilitate the appraisal of whether each of the nine instrument items is loaded onto respective factors as anticipated. The analysis displayed two factor loadings that were extracted with eigenvalues greater than 1 (Table 6.1). The first component had an eigenvalue of 4.67, while the second components had eigenvalue of 1.65. The two components accounted for 70% of the total variance. After the varimax rotation, the communicative health literacy items as well as the critical health literacy items all loaded onto factor 1, whereas the functional literacy items loaded onto the second factor. Concerning construct validity, the association of subscale scores to each other was evaluated. The scores on communicative health literacy

Table 6.1 Instrument item scores and principal component analysis factor loadings (infectious disease, James Town)

Instrument items	Item scores (means and SD)	Principal component analysis Factor loadings Factor 1	 Factor 2
How often do you need someone to help you when you are given information on gonorrhea to read by your doctor, nurse, or pharmacist?	1.71 (1.027)	0.203	0.794
When you need help to understand information about gonorrhea, can you easily get someone to assist you?	1.94 (1.07)	0.251	0.726
Do you need help to fill in official documents at the hospital or clinic?	2.26 (1.245)	−0.175	0.76
When you talk to a doctor or nurse about gonorrhea, do you give them all the information they need to help you?	1.70 (1.085)	0.873	0.185
When you talk to a doctor or nurse about gonorrhea, do you ask all the questions you need to ask?	1.71 (1.096)	0.874	0.163
When you talk to a doctor or nurse about gonorrhea, do you make sure they explain anything that you do not understand	1.69 (1.074)	0.907	0.147
Are you someone who likes to find out lots of different information about your health regarding gonorrhea?	1.52 (0.939)	0.82	−0.008
How often do you think carefully about whether health information on gonorrhea makes sense in a particular situation?	1.55 (0.902)	0.85	−0.028
Are you the sort of person who might question your doctor or nurse's advice on gonorrhea based on your own research/findings?	1.56 (0.952)	0.785	0.121

items were significantly correlated with functional health literacy score items ($r = 0.238$, $p = 0.003$) as well as the critical health literacy items ($r = 0.720$, $p < 0.001$). However, a statistically significant association was not found between functional health literacy scores and critical health literacy scores. Therefore, for total health literacy scores on infectious diseases, we used the six items measuring communicative and critical health literacy scores.

Bivariate Analyses: Health Literacy Scores and Demographic Characteristics

The demographic variables that were significantly associated with health literacy on gonorrhea in James Town include age, educational attainment, marital status, and occupation (Table 6.2). Age was negatively correlated with total health literacy on infectious disease and critical health literacy scores. Educational level was significantly associated with total health literacy on infectious disease and all the subscales—functional, communicative, and critical health literacy. Higher education was associated with higher mean ranked scores on health literacy. Consistently, those of tertiary education have higher scores on total health literacy and the subscales compared to the other categories. Also, migrants with high school education consistently had higher mean scores on health literacy compared to those with junior high education, primary education, and no formal education. Marital status was significantly related to total health literacy, interactive as well as critical health literacy on infectious disease. Specifically, married respondents had higher mean ranked scores on total health literacy than migrants who were divorced or separated although there was no significant difference among married migrants and their single counterparts. Single migrants scored higher than those who were divorced or separated. The same pattern is observed for the scores on interactive and critical health literacy on infectious disease. Occupation was significantly associated with total health

Table 6.2 Bivariate analyses: health literacy scores and demographic characteristics (infectious disease, James Town)

	Total health literacy—infectious disease	Functional health literacy—infectious disease	Communicative health literacy—infectious disease	Critical health literacy—infectious disease
Age	−0.188*	0.153	0.148	−0.207*
Education level	8.033***	5.124***	5.624***	8.597***
Marital status	5.118***	0.39	4.698***	4.925***
Occupation	3.447*	2.901*	1.78	5.047***

*$p < 0.05$, **$p < 0.01$, ***$p < 0.001$

literacy, functional as well as critical health literacies. Migrants in professional/managerial jobs had significantly higher scores in total health literacy, as well as in functional and critical health literacy than migrants in manual and sales jobs.

Brief Deliberations on Health Literacy on Gonorrhea in James Town

Generally, the result show low levels of functional, communicative, and critical health literacy among the study respondents. Almost two-thirds of the respondents reported not to provide medical staff with all the information needed to help them. Although STDs rank very high among the cases reported in the clinic (see Chap. 3), we speculate that, due to the sensitive nature of the subject, migrants may be uncomfortable discussing much details even with medical staff who are supposed to help them. This may be an entry point for health professionals across the spectrum of the healthcare ladder—local clinic to teaching hospitals—to be given professional development training regularly on how to make clients and patients comfortable so that discussion about STDs are had in a relaxed atmosphere. Like many STDs, gonorrhea has a social stigma and therefore may not be a subject matter people will readily open up to discuss with a healthcare professional. The role of healthcare professionals in promoting health literacy among the poor in this slum may be as important as general literacy, especially given that three-quarters of the migrants surveyed do not find out information about STDs and another three-quarters never give thought to health information on STDs making sense.

Sociodemographic variables explained the differences in health literacy in the slum, despite the general low health literacy levels observed. Age, education, marital status, and occupation were all significantly associated with health literacy. Increasing age was associated with decreasing score on total health literacy and critical health literacy. We are not sure why that relationship and that may warrant further investigation. Education was positively associated with total health literacy, functional, interactive, and critical health literacy. That is, people with higher education are more likely to, among other things, provide medical staff with all the information they need to help them, ask medical professional all the questions they feel they need to ask, ensure health professionals explain anything they do not understand, and find out lots of different information about their health. Marital status was significantly related with total health literacy and all the subscales. Specifically, married migrants were more likely to provide medical staff with all the information they need to help them, ask medical professional all the questions they feel they need to ask, ensure health professionals explain anything they do not understand, and find out lots of different information about their health than their divorce or separated counterparts. Single migrants scored better than their divorced/separated counterparts in total health literacy, functional health literacy, communicative, and critical health literacy.

Results on Health Literacy on Common Cold in Old Fadama

Descriptive Statistics

As previously indicated, the infectious disease chosen for evaluation in Old Fadama was an upper respiratory disease (common cold). Regarding functional literacy, more than 50% of the respondents reported that they need help when they are given information on common cold to read by a doctor, nurse, or pharmacist while 33% reported not needing help at all. Forty and 20% reported that they can easily get help to comprehend information on common cold "sometimes" and "often," respectively. On communicative health literacy, while about 30% indicated that they do not always provide the nurse or doctor all the information they need when they talk to them about common cold, about 36 and 14% reported to "sometimes" and "often" ask all the questions they need to ask the doctor or nurse, respectively, when they talk to them. Concerning critical health literacy, when asked whether they like to find out lots of different information about your health regarding common cold, 51% indicated "not at all," while 13 and 2% indicated "sometimes" and "often," respectively. When asked about how often they thought carefully about whether health information on common cold made sense in a particular situation, 50% indicated "not at all," while 19 and 2% reported "sometimes" and "often," respectively.

Reliability Analyses

To investigate the internal consistency of the main items of the scale, reliability analyses were performed. A very good Cronbach alpha value of 0.828 was achieved for the total scale. Concerning each of the subscales, the Cronbach alpha value for the items evaluating functional health literacy on infectious disease was adequate (0.630), whereas the Cronbach alpha value for the three items appraising communicative health literacy was excellent (0.907). The analyses on the measure for the three critical health literacy items reached a very satisfactory value of 0.742.

Content and Construct Validity: Principal Component Analysis and Correlations

To explore content validity regarding the fundamental structure of the nine items on the scale, a principal component analysis with varimax rotation was executed. Given that the items were assessing the three health literacy typologies, this analysis was to facilitate the assessment of whether each of the nine instrument items is loaded onto respective factors as estimated. The analysis exhibited two factor loadings that were extracted with eigenvalues greater than 1 (Table 6.3). The first component had an eigenvalue of

Table 6.3 Instrument item scores and principal component analysis factor loadings (infectious disease, Old Fadama)

Instrument items	Item scores (means and SD)	Principal component analysis Factor loadings Factor 1	Factor 2
How often do you need someone to help you when you are given information on cold to read by your doctor, nurse, or pharmacist?	2.46 (1.196)	0.235	0.794
When you need help to understand information about cold, can you easily get someone to assist you?	2.60 (1.043)	−0.026	0.741
Do you need help to fill in official documents at the hospital or clinic?	3.16 (1.142)	0.114	0.645
When you talk to a doctor or nurse about cold, do you give them all the information they need to help you?	2.56 (1.179)	0.724	0.447
When you talk to a doctor or nurse about cold, do you ask all the questions you need to ask?	2.30 (1.085)	0.76	0.416
When you talk to a doctor or nurse about cold, do you make sure they explain anything that you do not understand	2.13 (1.079)	0.773	0.282
Are you someone who likes to find out lots of different information about your health regarding cold?	1.66 (0.784)	0.762	0.025
How often do you think carefully about whether health information on cold makes sense in a particular situation?	1.73 (0.841)	0.77	0.03
Are you the sort of person who might question your doctor or nurse's advice on cold based on your own research/findings?	1.51 (0.775)	0.65	−0.021

4.01, while the second components had eigenvalue of 1.40. The two components accounted for 60% of the total variance. After the varimax rotation, the communicative health literacy items as well as the critical health literacy items all loaded onto factor 1, while the functional literacy items loaded onto factor 2. About construct validity, the connection between subscale scores to each other was assessed. The scores on communicative health literacy items were significantly correlated with functional health literacy score items ($r = 0.429$, $p < 0.001$) as well as the critical health literacy items ($r = 0.567$, $p < 0.001$). Also, a statistically significant association was found between functional health literacy scores and critical health literacy scores ($r = 0.248$, $p < 0.05$).

Bivariate Analyses: Health Literacy Scores and Demographic Characteristics

The demographic variables that were significantly related with health literacy on common cold in Old Fadama include educational attainment, marital status, occupation, and household size (Table 6.4). Educational level was significantly associated

Table 6.4 Bivariate analyses: health literacy scores and demographic characteristics (infectious disease, Old Fadama)

	Total health literacy—infectious disease	Functional health literacy—infectious disease	Communicative health literacy—infectious disease	Critical health literacy—infectious disease
Education level	3.788*	2.211	3.177*	2.901*
Marital status	2.380	1.260	1.719	3.606*
Occupation	1.833	1.913	3.003*	0.536
Household size	0.158	0.018	0.109	0.192*

*$p < 0.05$

with total health literacy on infectious disease and two of the subscales—communicative and critical health literacy. Higher education was associated with higher mean ranked scores on health literacy. Constantly, those with secondary school and higher education had significantly higher scores on total health literacy and the two subscales compared to migrants with junior high education, primary education, and no formal education. Marital status was significantly related to critical health literacy on infectious disease (common cold). Precisely, single migrant respondents had higher mean rank scores on critical health literacy on common cold than migrants who were divorced or separated and cohabiting. There was no significant difference among married migrants and other categories although married migrants had higher mean rank scores. Occupation was significantly associated with communicative health literacy. Migrants in professional/managerial jobs, skilled manual, and sales jobs had significantly higher scores in communicative health literacy than migrants in unskilled manual jobs. Household size was positively correlated with critical health literacy.

Brief Deliberation on Health Literacy on Common Cold in Old Fadama

Health literacy on common cold among migrants in Old Fadama is generally low. While the lower levels are more evident in the scores on functional health literacy, there seem to be higher score on communicative health literacy. About 70% of migrants reported that they provide health professionals with the needed information to help them. We speculate that common cold is not associated with social stigma compared to STDs like gonorrhea and that may be responsible for the higher score on that communicative health literacy variable. Approximately half of the respondents reported that they deliberately seek information about their health regarding common cold.

Health literacy on common cold in Old Fadama is demographically differentiated. Education was significantly associated with total health literacy, communicative health literacy, and critical health literacy. That means, overall, migrants with higher education consistently scored better on ability to read and comprehend information given to them by medical staff, complete official documents at the hospital or clinic without assistance, provision of the necessary information to a doctor or nurse or pharmacist, and making medical personnel explain issues they do not understand. Regarding marital status, although there was no significant difference between the health literacy scores of single migrants and their married counterparts, single migrants scored higher on critical health literacy including subjects such as skills pertaining to questioning one's doctor or nurse's advice on cold based on their own research/findings and willingly seeking information about their health as it relates to common cold.

Health Literacy on Noncommunicable Diseases Among Migrants

The data from the community clinics mainly led to our choices of noncommunicable disease to investigate. For both James Town and Old Fadama, the leading noncommunicable disease reported was hypertension and educational programs on it have been undertaken in the communities. Like the design for the health literacy on infectious diseases assessment, a nine self-report items premised on Nutbeam (2000) model was developed. That is, the dimensions of health literacy that were examined are functional, communicative/interactive, and critical health literacy. The core constructs for each of the typologies as discussed for infectious diseases remain the same. There were three items per construct. Each of the nine items had the following response options and codes: "not at all,"—1; "rarely,"—2; "sometimes,"—3; and "often"—4.

Analyses

In analyzing the data on health literacy on noncommunicable diseases, we employed descriptive statistics, reliability analyses, validity analyses, independent samples *t*-test, Pearson correlation, and analysis of variance (ANOVA). A description of the percentage distribution of the responses to the nine items measuring functional health literacy, interactive/communicative health literacy, and critical health literacy is discussed. An appraisal of the internal consistency of the scale of the whole instrument as well as the subscales was undertaken through calculations of Cronbach alpha coefficients. Subsequently, a principal component analysis using varimax rotation was carried out to assess the factor structure of the scale. We computed health literacy scores for the whole scale and subscales and successively measured the relationship between health literacy and demographic characteristics using ANOVA, Pearson correlation, and independent samples *t*-test.

Results on Health Literacy on Hypertension in James Town

Descriptive Statistics

About functional health literacy on hypertension, while about 50% of respondents indicated that they do not at all need help when they are given information on hypertension to read by a doctor, nurse, or pharmacist, 27 and 13% reported to need help "sometimes" and "often," respectively. Twenty-eight and 12% reported that they can easily get help to understand information on hypertension "sometimes" and "often," respectively. Regarding interactive or communicative health literacy on hypertension, 23 and 19% of migrant respondents asserted they "sometimes" and "often," respectively, provide the nurse or doctor or pharmacist all the information they need when they talk with them, whereas 52% indicated "not at all." Meanwhile, only 21 and 14% of the migrants interviewed reported to "sometimes" and "often," respectively, ask the nurse or doctor all the questions they need to ask. Regarding critical health literacy, when asked whether they like to find out lots of different information about their health regarding hypertension, 56% indicated "not at all," while 25 and 9% indicated "sometimes" and "often," respectively. When asked about how often they thought carefully about whether health information on hypertension made sense in a particular situation, 56% indicated "not at all," while 30 and 5% reported "sometimes" and "often," respectively.

Reliability Analyses

Reliability analyses through Cronbach alpha calculations to ascertain the internal consistency of the main items of the scale were performed. A very good Cronbach alpha value of 0.880 was achieved for the total scale. Relating to each of the subscales, the Cronbach alpha value for the items evaluating functional health literacy on hypertension was very satisfactory (0.710), while the Cronbach alpha value for the three items assessing interactive health literacy on hypertension was excellent (0.916). The three critical health literacy items attained an excellent Cronbach alpha value of 0.899.

Content and Construct Validity: Principal Component Analysis and Correlations

A principal component analysis with varimax rotation was undertaken to explore content validity about the fundamental structure of the nine items on the scale. Bearing in mind that the nine items were assessing the three health literacy typologies, with three items per typology, this analysis was to enable the measurement of whether each of the nine instrument items is loaded onto their respective factors as

Table 6.5 Instrument item scores and principal component analysis factor loadings (noncommunicable disease, James Town)

Instrument items	Item scores (means and SD)	Principal component analysis Factor loadings	
		Factor 1	Factor 2
How often do you need someone to help you when you are given information on hypertension to read by your doctor, nurse, or pharmacist?	2.01 (1.135)	0.125	0.842
When you need help to understand information about hypertension, can you easily get someone to assist you?	2.02 (1.126)	0.346	0.626
Do you need help to fill in official documents at the hospital or clinic?	2.35 (1.264)	−0.008	0.851
When you talk to a doctor or nurse about hypertension, do you give them all the information they need to help you?	2.10 (1.236)	0.763	0.381
When you talk to a doctor or nurse about hypertension, do you ask all the questions you need to ask?	1.92 (1.156)	0.84	0.224
When you talk to a doctor or nurse about hypertension, do you make sure they explain anything that you do not understand?	2.11 (1.218)	0.897	0.206
Are you someone who likes to find out lots of different information about your health regarding hypertension?	1.87 (1.072)	0.81	0.13
How often do you think carefully about whether health information on hypertension makes sense in a particular situation?	1.85 (1.028)	0.886	0.08
Are you the sort of person who might question your doctor or nurse's advice on hypertension based on your own research/findings?	1.75 (0.998)	0.881	−0.015

projected. Results indicated that two factor loadings were extracted with eigenvalues greater than 1 (Table 6.5). The first component had an eigenvalue of 4.90, while the second components had eigenvalue of 1.62. The two components accounted for 73% of the total variance. After the varimax rotation, the communicative health literacy items as well as the critical health literacy items all loaded onto factor 1, while the functional literacy items loaded onto factor 2. Consequently, the total health literacy score was computed using communicative and critical health literacy scores. Regarding construct validity, the correlation between subscale scores to each other was evaluated. The scores on communicative health literacy items were significantly correlated with functional health literacy score items ($r = 0.423$, $p < 0.001$) as well as the critical health literacy items ($r = 0.766$, $p < 0.001$). Also, a statistically significant association was found between functional health literacy scores and critical health literacy scores ($r = 0.268$, $p < 0.001$).

Table 6.6 Bivariate analyses: health literacy scores and demographic characteristics (noncommunicable disease (NCD), James Town)

	Total health literacy—NCD	Functional health literacy—NCD	Communicative health literacy—NCD	Critical health literacy—NCD
Sex	3.207*	2.335**	2.163**	1.161
Age	0.228*	0.228*	0.239*	0.081
Employment status	−0.785	−0.754	1.454	2.003**
Education level	6.440***	6.445***	0.946	1.941

*$p < 0.01$, **$p < 0.05$, ***$p < 0.001$

Bivariate Analyses: Health Literacy Scores and Demographic Characteristics

In James Town, the demographic variables found to be associated with health literacy on hypertension were sex, age, employment status, and educational attainment (Table 6.6). Sex was significantly associated with total health literacy, functional health literacy, as well as critical health literacy. Females had significantly higher mean scores than males in all the health literacy typologies. Concerning age, migrants' age was positively correlated with total health literacy score, functional health literacy, and communicative health literacy. Regarding employment status, surprisingly, migrant respondents who were unemployed had higher means scores on critical health literacy compared with those who were employed. Education was significantly associated with total health literacy score and functional health literacy. As expected, higher educational attainment was associated with better functional literacy score and higher total health literacy score.

Old Fadama

Descriptive Statistics

As regards functional health literacy on hypertension in Old Fadama, while about 41% of respondents indicated that they do not at all need help when they are given information on hypertension to read by a doctor, nurse, or pharmacist, 33 and 19% reported to need help "sometimes" and "often," respectively. About 39 and 18% reported that they can easily get help to understand information on hypertension "sometimes" and "often," respectively. Considering interactive or communicative health literacy on hypertension, though 21 and 19% of migrant respondents asserted they "sometimes" and "often," respectively, provide the nurse or doctor or pharmacist all the information they need when they talk with them, 44% indicated that they do not at all provide health professional with all the information they need. Meanwhile, only 21 and 15% of the migrants interviewed reported to "sometimes" and "often," respectively, ask the nurse or doctor all the questions they need to ask. Relating to critical health literacy, when asked whether they like to find out lots of different information about their health regarding hypertension, while 16 and 1% indicated "sometimes" and "often," respectively, 49% reported to not at all like to do so. When asked about how often they thought carefully about whether health information on hypertension made sense in a particular situation, 61% indicated "not at all," while 11 and 1% reported "sometimes" and "often," respectively.

Reliability Analyses

We computed Cronbach alpha coefficients to determine the internal consistency of the nine items of the scale measuring functional, communicative, and critical health literacy on hypertension in Old Fadama. For the total scale, a very good Cronbach alpha value of 0.821 was achieved. Concerning each of the subscales, the Cronbach alpha value for the items evaluating functional health literacy on hypertension was very satisfactory (0.701), while the Cronbach alpha value for the three items measuring communicative health literacy on hypertension was excellent (0.929). The three critical health literacy items attained an excellent Cronbach alpha value of 0.767.

Content and Construct Validity: Principal Component Analysis and Correlations

A principal component analysis with varimax rotation was used to investigate content validity about the fundamental structure of the nine items on the scale measuring health literacy. Because the nine items were evaluating the three health literacy typologies, with three items per typology, this analysis was to enable the assessment of whether each of the nine instrument items is loaded onto their individual factors as anticipated. We found that two factor loadings were extracted with eigenvalues greater than 1 (Table 6.7). The first component had an eigenvalue of 3.96, while the second components had eigenvalue of 1.75. The two components accounted for 64% of the total variance. Following the varimax rotation, the communicative health literacy items together with the critical health literacy items all loaded onto factor 1, whereas the functional literacy items loaded onto factor 2. Concerning construct validity, the correlation between subscale scores of each typology to each other was appraised. The scores on communicative health literacy items were significantly correlated with functional health literacy score items ($r = 0.391$, $p < 0.001$) as well as the critical health literacy items ($r = 0.533$, $p < 0.001$). However, a statistically significant association was not found between functional health literacy scores and critical health literacy scores. Therefore, the computation of the total health literacy score on hypertension utilized the communicative and critical health literacy scores.

Table 6.7 Instrument item scores and principal component analysis factor loadings (noncommunicable disease, Old Fadama)

Instrument items	Item scores (means and SD)	Principal component analysis Factor loadings	
		Factor 1	Factor 2
How often do you need someone to help you when you are given information on hypertension to read by your doctor, nurse, or pharmacist?	2.29 (1.190)	0.14	0.753
When you need help to understand information about hypertension, can you easily get someone to assist you?	2.27 (1.092)	0.067	0.807
Do you need help to fill in official documents at the hospital or clinic?	3.13 (1.166)	−0.04	0.702
When you talk to a doctor or nurse about hypertension, do you give them all the information they need to help you?	2.16 (1.188)	0.697	0.465
When you talk to a doctor or nurse about hypertension, do you ask all the questions you need to ask?	2.03 (1.135)	0.779	0.409
When you talk to a doctor or nurse about hypertension, do you make sure they explain anything that you do not understand?	1.93 (1.056)	0.758	0.45
Are you someone who likes to find out lots of different information about your health regarding hypertension?	1.69 (0.785)	0.763	−0.004
How often do you think carefully about whether health information on hypertension makes sense in a particular situation?	1.53 (0.739)	0.846	−0.142
Are you the sort of person who might question your doctor or nurse's advice on hypertension based on your own research/findings?	1.42 (0.616)	0.638	−0.039

Bivariate Analyses: Health Literacy Scores and Demographic Characteristics

In Old Fadama, the demographic variable found to be associated with health literacy on hypertension was household size. Specifically, household size was significantly associated with critical health literacy on hypertension but not functional or interactive health literacy. Higher household size was associated with better score on critical health literacy ($r = 0.165$, $p < 0.05$).

Brief Deliberation on Health Literacy on Hypertension in James Town and Old Fadama

The prevalence of hypertension in both slums, as discussed in Chap. 3, epitomizes the double burden of disease on the urban poor in Accra, Ghana. The investigation of health literacy in these slums reveals low levels of health literacy, and with the seeming increasing number of hypertension cases, poor health literacy may aggravate the plight of migrants. Although both slums have lower health literacy on hypertension, the scores show that Old Fadama migrants are worse off.

In James Town, health literacy was demographically differentiated with sex, age, and education being significantly related to health literacy. Sex was significantly related to total health literacy and all the subscales. Specifically, females scored higher than males on health literacy skills and competencies such as ability to complete hospital documents, capacity to provide health professionals with all necessary information about the health situation, ability to ask healthcare providers all mind-lingering questions when they interact with them, capability to seek explanation from medical personnel about issues they may not initially understand, as well as queries doctor's advice on hypertension through personal research or findings. Increasing age was associated with better total health literacy, functional and communicative health literacy on hypertension. These findings may be attributed to, all other things being equal, the fact that age varies directly or positively with hypertension diagnosis. Education had a positive association with total health literacy score and functional health literacy. The higher one's education, the more likely it is for the person to comprehend instructions by medical staff and not require help to read medication instructions.

References

Bann CM, McCormack LA, Berkman ND, Squiers LB (2012) The health literacy skills instrument: a 10-item short form. J Health Commun 17(suppl 3):191–202. https://doi.org/10.1080/10810730.2012.718042

Chinn D, McCarthy C (2013) All Aspects of Health Literacy Scale (AAHLS): developing a tool to measure functional, communicative and critical health literacy in primary healthcare settings. Patient Educ Couns 90(2):247–253. https://doi.org/10.1016/j.pec.2012.10.019

Chung SY, Nahm ES (2015) Testing reliability and validity of the eHealth Literacy Scale (eHEALS) for older adults recruited online. Comput Inform Nurs 33(4):150–156. https://doi.org/10.1097/cin.0000000000000146

Kim MT, Song HJ, Han HR, Song Y, Nam S, Nguyen TH et al (2012) Development and validation of the high blood pressure-focused health literacy scale. Patient Educ Couns 87(2):165–170. https://doi.org/10.1016/j.pec.2011.09.005

Nutbeam D (2000) Health literacy as a public health goal: a challenge for contemporary health education and communication strategies into the 21st century. Health Promot Int 15(3):259–267

Paasche-Orlow MK, Wolf MS (2007) The causal pathways linking health literacy to health outcomes. Am J Health Behav 31(1):S19–S26

Chapter 7
Social Capital and Acculturation

Abstract In this chapter, we discuss the interrelationship between social capital, acculturation, lifestyle, and health. First, we explore the forms of social capital acquired at the destination and the constituents of social capital formation. Secondly, an assessment of the impact of social relations in the acculturation process is undertaken. Further, we show how social relations influence lifestyles that border on diverse forms of acculturation inclinations including assimilation, integration, separation, and marginalization. Thirdly, an examination of the perceptions of the impact of lifestyles on health outcomes is undertaken. Conceptual mapping was used to elicit information from migrants. Acculturation elements enabled by social relations include language acquisitions and comprehension, food, customs and values, as well as clothing styles. Forms of social capital such as bonding, bridging, and linking social capital were present in the social capital formation of migrants and each construct contributed uniquely and jointly to acculturation preferences and inclinations. For example, the place (community) and the ability to function effectively as a member of the community influenced one's acculturation inclination regarding language. Migrants indicated that while some lifestyles learned from social relations, as adaptive strategies, may have positive effects on health outcomes, other lifestyles were mentioned as having detrimental health effects.

Keywords Acculturation · Integration · Assimilation · Social capital · Social relations · Lifestyles · Conceptual mapping · Adaptive strategies · Migrants' health

Introduction

This chapter explains the complexities of the social capital concept and illustrates how its constructs enable acculturation process at the place of destination among migrants. Acculturation inclinations such as integration, marginalization, separation, and assimilation are explained to understand their relations with social capital

R. A. Tutu, J. D. Busingye, *Migration, Social Capital, and Health*,
Global Perspectives on Health Geography,
https://doi.org/10.1007/978-3-030-24693-8_7

formation and use. We describe how different forms of social relations influence various lifestyles and behavior with implications for health.

Acculturation debates are over four decades old. It refers to the extent to which migrants voluntarily opt to be involved in the host society relative to holding on to their native cultural identity (Berry et al. 2006). Acculturation strategies have been classified into four. These are integration, assimilation, separation, and marginalization. Integration as an acculturation strategy refers to a situation when migrants decide to maintain both their original cultural identity and that of the host population at the same time. Assimilation occurs when migrants discard their original cultural practices and wholly adapts the practices, norms, and values of the host population. Separation happens when migrants continue their original cultural practices at the destination and shuns interaction with the host population and their cultural practices. Finally, when migrants show little interest in the maintenance of their own cultural values as well as the practices of the host population at the place of destination, marginalization is said to have occurred.

A migrant's acculturation inclination is dependent on several factors. These include the previous experiences of migrants—the reasons for moving from the place of origin, socioeconomic status in their place of origin, previous migration experience, age at migration, and educational attainment (Braun and Glöckner-Rist 2012; Kalmijn and Van Tubergen 2006; Mondain and Lardoux 2013); generation effects—the youth tend to adjust with and adapt new and different values compared to their parents (Mondain and Lardoux 2013); participation spaces—where they will be able to partake in their original customs, values, and cultural practices like food, religious observance, and baby naming ceremonies; as well as one's duration of stay at the destination—because length of stay augments the attainment of knowledge of the cultural dynamics at the destination while enabling gaining of practices and language essential for economic engagement (Djamba and Kimuna 2012). The duration of stay at the destination by a migrant and its relationship to acculturation is mediated by attitudinal changes. According Frank and Heuveline (2005), longer duration of stay among the host population influences attitudinal changes, which leads to migrants becoming more accepting of the cultural practices of the host population.

Social capital is instrumental in determining acculturation inclinations as well as in shaping the acculturation process. Migrants loose social capital when they leave their homes and place of origins. This is because they leave behind social relations that may help them acquire resources (the "capital" in social capital) through, among other things, child care, job search, emotional support, and provision of shelter (Nakhaie and Kazemipur 2013). Meanwhile, social capital type formation at the destination upon arrival may not help with job acquisition, and even if it does, it does not necessarily contribute to job upward mobility (Nakhaie and Kazemipur 2013) and the jobs may be less prestigious (Sanders et al. 2002).

That notwithstanding, social capital may be gained when the varying forms of social capital are examined. Three different forms of social capital of relevance to this discussion are bonding, bridging, and linking social capital (Hawkins and Maurer 2010). Bonding social capital denotes strong parallel connections between persons or groups sharing comparable sociodemographic characteristics. These include strong ties with family members and immediate relations. Bonding social capital has been found to have a positive impact on acculturation. Bridging social capital symbolizes relations outside one's immediate family, but traverse diverse people. This kind of social capital, usually referred to as weak ties, includes friends and group relations such as religious organizations. Linking social capital, however, represents upright social relations—people having ties with individuals wielding social power over them.

The optimal interplay of these forms of social capital grants migrant's ability to adapt and meet different goals at the destination. For instance, in accessing resources at the destination, forms of social capital interchange to ensure a level of equilibrium depending on how strong or weak a tie is at the destination. In a study on social capital and employment in Canada, Nakhaie and Kazemipur (2013) found that the strength of ethnic homogeneity ties determined the extent to which migrants obtained employment opportunities and high rank work. They found bridging and linking social capital being higher when friendship networks based on ethnic homogeneity was weak.

Raza et al. (2013) centralize social capital in acculturation through the modification of Berry's acculturation model and demonstrate the inextricable link between the two separate but connected concepts. The authors used the idea of weak or strong forms of social capital to illustrate the relationship. The notion of "individualization" suggested by Raza et al. (2013) refers to the occurrence of weak bonding as well as weak bridging social networks among migrants, which is a product of cultural seclusion or reproof of one's own cultural practices. Consequently, "individualization" is the epitome of the marginalization construct of acculturation. Further, by that same lucidity, the presence of strong bonding and strong bridging social capital demonstrates integration, whereas strong bridging and weak bonding social capital depicts separation as an acculturation strategy. The manifestation of strong bridging and weak bonding social capital is expressive of assimilation strategy.

These interconnected social ties and acculturation strategies influence lifestyles with health implications. For example, migrant youth usually form groups, which may be based on ethnic lines or age in the context of the Global South. Such relations are premised on adherence to group practices and norms through regular interactions with each other. Therefore, principles governing this kind of relationship—commitment, reciprocity, and equal control—are adhered vigorously to ensure acceptance (Teja and Schonert-Reichl 2013). Consequently, the desire to fit in and to please members of the group may have positive and/or negative facets. Although friends provide camaraderie, assistance, intimacy, and facilitate quarrel resolution, some negative components are conflicts, poor friendship quality that may lead to behavioral difficulties, as well as disloyalty (Teja and Schonert-Reichl 2013).

Next, we present findings on the relationship between social capital, acculturation, lifestyle, and health from our study sites. We employed conceptual mapping (mental models) as the method of data collection. Subsequently, we discuss the main findings and tease out the role of place in these dynamics.

Conceptual Mapping on Social Capital, Acculturation, Lifestyle, and Health

Mental model activity was undertaken to help us understand how residents think about social capital formation and its impact on acculturation and lifestyles, which subsequently affects health outcomes. Social capital was central in our thinking and that was the starting point of the activity. Social capital—actual resources (e.g., money, clothing, etc.) flow from social relations to individuals—was written boldly in the middle of a large piece of paper. Whenever needed, a symbol (resource/person, etc.) of social capital, requested by the participant, was drawn on a separate sheet next to the main sheet for the activity.

Participants were asked to brainstorm about all the forms of social relations (e.g., friends, brothers, etc.) they have from which material resources are gained. We probed for proximity of relations across space. Each of the forms of the social relations identified by participants was written on sticky notes. The kind of relations were connected to each other and to social capital with arrows based on the responses and discussions that ensued.

Participants were asked to discuss how these social relations in the community and city help them to adapt to the city upon arrival. We probed for elements of acculturation (language, food, city life, city customs and values, coping with anxiety, etc.) whenever they were not mentioned. Each of the acculturation/adaptation components mentioned by participants was written on sticky notes and connected with arrows to each other and to social capital whenever applicable.

Subsequently, migrants were asked to brainstorm about how social relations influence lifestyles (behavior and practices) generally. In this case, it was based on their general observation in the community and not necessarily in their own case. However, respondents were encouraged to share their personal experiences whenever they desire. Specific lifestyles were written on the sticky notes and connected to each other and to social capital using arrows based on the respondents' views on the myriad interconnections between social relations and lifestyles. Finally, migrants were asked to suggest how each lifestyle they have identified affects health outcomes/diseases/illnesses. That is, the health consequences of the lifestyles identified. Each identified consequence was also written on sticky notes and connection made with arrows to lifestyles acknowledged earlier.

Results from James Town Activities

Social Relations and Acculturation

Social relations that contribute to the material needs of migrants as mentioned include family members—father, mother, siblings, mother-in-law, and spouses; religious organizations—church and church members; associations and clubs; and linking social capital—politicians and opinion leaders. These social relations serve as a conduit for diverse forms of acculturation among migrants. The main elements of acculturation, which social relations aid in adapting, indicated by migrants include language learning, food, customs and values, and clothing.

Learning the Ga language, which is the indigenous language spoken by the natives of James Town, was of paramount importance to the migrants. They explained how that enables them to freely interact with their host community, understand what is going on in their immediate neighborhoods, and make friends with their host. Inability to speak or understand Ga seems to put migrants at a disadvantage regarding acculturation, according to our respondents. Consequently, newly arrived migrants make friends with the host population to learn Ga or their family members who understand Ga serve as their teachers as well. A participant retorted that:

> I have learned to understand the language of the indigenous people [Ga] and this was facilitated by friends [in the community]. I also eat the staple food through my own efforts. But I cannot prepare that staple dish. [Female; 35 years; 10 year in community]

The importance of the language learning notwithstanding, while a respondent acknowledged language utility for interaction, he emphasized his satisfaction with the local food and disinterest in learning the language. He said:

> For the food [of the Ga people], it has been so easy for me to eat the food in this community. But for the language, I am reserved. I have not learned any of their languages [dialects] so far. [Male; 55 years; 14 years in community]

Regarding acculturation in the area of food, the primary elements mentioned include the types of food, cooking patterns, and perceptions of the extent of healthy nature of food. The changing of the type of food to consume is facilitated primarily by family and friends. A female respondent explained how social relations enabled her to adapt for the staple food in the host community as follows:

> There has been a change in the type of food I eat since I came to live in this community. Through the help of my mother-in-law and husband, I can now eat certain foods I never knew how to eat when I first came to live here. [Female; 21 years; 2 years in community]

The adaptation to the new environment staple foods also impacts cooking patterns. The respondent quoted above explained how the city life and familiarity with the indigenous food has affected her cooking pattern:

> Also, there has been a change in my pattern of cooking. When I was in my home town, I used to cook all the time; that is, from Monday to Sunday and this is our way of life. But since I came to live in this community, it has changed. I now even cook once or twice a week. [This] is due to the fact my husband does not come home early from work and due [to that] I always buy food from outside or street vendors and eat with my little daughter most of the time. [Female; 21 years; 2 years in community]

Some respondents perceived the changing patterns of their food consumption are health and allowing for variety, which was not available to them. Therefore, although one may perceive the place as a slum, the available food presents opportunities. A respondent indicated that:

> I have adapted to city life especially the food I eat. I eat a lot of vegetables because they are common here compared to where I used to live. I also eat other type of fishes and meat that are accessible here but not my hometown where I come from. [Female; 40 years; 15 years in community]

Concerning customs, mention was made of the traditional and customary practices of the host population. Such practices include traditional funeral rites, baby naming ceremonies, and twin's celebration. The responses to the extent of acculturation on these customs were varied. While a couple of respondents felt normal and fascinated participating in these customs, one migrant taught it was against his religious beliefs and hence could not fully engage those practices.

> …For their customs and values, I am not malleable to it especially the religious practice in this community is more of the traditional religion among the indigenous groups and not [goes contrary to] mine. [Male; 55 years; 14 years in community]

A migrant admires and has adapted the customary processes in the preparation of social events. Reference is made to the robust and meticulous nature of the adherence to hierarchical order and processes for organizing such social events. A female respondent indicated the following:

> One thing I have pick from here is the customs and values in the preparation towards social event such as funeral rites… [Female; 35 years; 10 years in community]

On clothing, the question of decency was primary on the minds of the participants. While one migrant mentioned the customary attire of the indigenous host population as admirable, most of the discussion of other respondents centered around the decency or lack thereof among residents. One respondent indicated that:

> You see some of the migrants dressing indecently. But other members of the community also dress decently. I mean not being half-way naked… This is usually [learned] from community members. [Female; 32 years; 13 years in community]

Lifestyles and Health

Lifestyle, which was explained as how or the way people live their lives in the community, indicated by migrants include carefree living, alcoholism, quarrelsome attitudes, fighting, drug usage, indecent dressing, poor eating habits, and indiscriminate disposal of waste. These lifestyles are intertwined in many respects and are perceived to be responsible for health outcomes such as infectious diseases, high fever, liver problems, hypertension, and malaria as well as injuries, spiritual attacks, and death. Regarding carefree living, two main opinions emerged: (1) the perception that carefree life is responsible for the indiscriminate dumping of refuse in the community and child delinquency, and (2) carefree lifestyle relates to the penchant for fighting among residents. Firstly, the carefree attitudes resulting in dumping of garbage in drains lead to choked gutters, which become breeding grounds for mosquitoes thereby leading to increased malaria cases in the community. This interconnection by migrants is epitomized in a couple of quotes below:

> Community members do things anyhow in terms of their sanitation, they throw refuse anywhere and anyhow they like, and this has led to rampant communicable diseases like malaria in the community for which we are all at risk [Male; 55 years; 14 years in community]

> …dumping of refuse anyhow and in the gutters also breed mosquitoes. Choked gutters which mostly breeds mosquitoes leading to the contracting of the malaria parasite. [Female; 21 years; 2 years in community]

Child delinquency was attributed to inadequate parental and guardian supervision in the community. Therefore, children engage in behavior that make them susceptible to negative health outcomes. This is articulated by a respondent as follows:

> Child delinquency especially among migrant children has led to the contracting of health-related issues like injuries; even including malaria which is also communicable [because they] sleep outside [in the open] sometimes. [Male; 55 years; 14 years in community]

Secondly, the penchant for fighting is suggested to be a trace picked-up from the host community, which results in injury and death. Fighting was linked to alcoholism and quarrelsome behaviors. Respondents recounted incidence of individual fist fighting that degenerate into households fighting against each other. A migrant considered the extent of fighting and its frequency in the community to be excessive and the ease with which migrants may emulate:

> There is too much noise making and fighting in this community and it is common among the indigenes. It is easy for one to pick up such behaviors. These too much fighting sometimes results in injury and even death do occur later. [Female; 35 years; 10 year in community]

Interesting connections were found among decency, spirituality, and health outcomes. While the relationship between indecent dressing and promiscuity seemed prominent in the responses of lifestyle, what we did not expect was the link to spirituality. Specifically, a link was made between indecent dressing and exposure to the displeasure of spiritual deities thereby incurring their wrath. A participant put it this way:

> Indecent way of dressing which usually makes community members experience spiritual attacks all the time. It is so serious that they can even be infected with cancer spiritually. [Female; 21 years; 2 years in community]

The health professional participants emphasized poor eating habits, poor nutrition, and substance use as lifestyles that most detrimental to the health of migrants. Concerning poor eating habits and nutrition, one health professional mentioned the poor nutritional value in some of the staple foods in the community. Inadequate balanced diet contributes to various forms of malnutrition problems among migrants and the host population as well. Furthermore, the health professional indicated that the cases of diseases related to the internal organs of migrants in the community, in their opinion, are linked to substance abuse. One of them remarked:

> Other lifestyle to include learning to smoke substances, especially among migrants in this community. This is usually [copied] from community members. The lifestyle of alcoholism and use of substance [marijuana] could result in diseases such as damage to the liver, kidney failure, and hypertension [which] we are seeing. [Female; 32 years; 13 years in community]

The unbalanced diet and poor nutrition were admitted to by a respondent as a personal poor lifestyle and proceeded with her mental depiction of other lifestyles and their implications for health. She talked about processed noodles (*Indomie*), which is available as street vendor food or purchased from a shop, easy to prepare, and how unhealthy such a food can be.

> One lifestyle that I have adapted that I know is not very good is the eating of unhealthy food such as only rice and *indomie* foods. But there are lifestyles that are not good in this community that one could easily adapt to such as having no respect for anyone. Alcoholism and drug [marijuana] usage is common here and this could result in high fever and mental problem. [Female; 24 years; 8 years in community]

We depict the connections made by migrants on their individual conceptual maps into an amalgamated mental model. Please see Figs. 7.1 and 7.2.

Results from Old Fadama Activities

Social Relations and Acculturation

The "capital" in social capital in Old Fadama according to our participants emanates from social relations that include nuclear family members—father and siblings; extended family members—uncle; friends; neighbors; and banks. The extent of their acculturation is a result of the social relations advice and admonishing. Elements of acculturation identified are language, food, work and business, as well as hygiene practices.

Regarding language, in Old Fadama, it was more about familiarity with the Twi language and less of the Ga language. Language is perceived by migrants as integral to their lifestyle in the city and crucial for how they interact in the neighborhood as well as in their daily work activities. Although, as previously noted, the community is divided along ethnic lines and people of same ethnicity have different enclaves, in the city, one of the commonest languages is Twi. While social relations at the destination helped them to learn and better understand Twi, others spoke and understand Twi prior to migration and therefore language was not a big communication barrier upon arrival. A couple of respondents who did not speak or understand Twi prior to migration but have been aided by friends and neighbor to speak said:

> Through the advice, teaching, and communication of my four friends in the Twi language, I can now speak the Twi language which I was unable to speak when I was in my hometown before [coming] here. [Male; 36 years; 7 years in community]

> I can now speak the Twi language through the help of my neighbors especially those who speak the Twi language and come from the places where the Twi language is spoken [Male; 44 years; 21 years in community]

Regarding food, participants mentioned types of food and eateries as main acculturation elements. While a participant acknowledged that migration into the city has

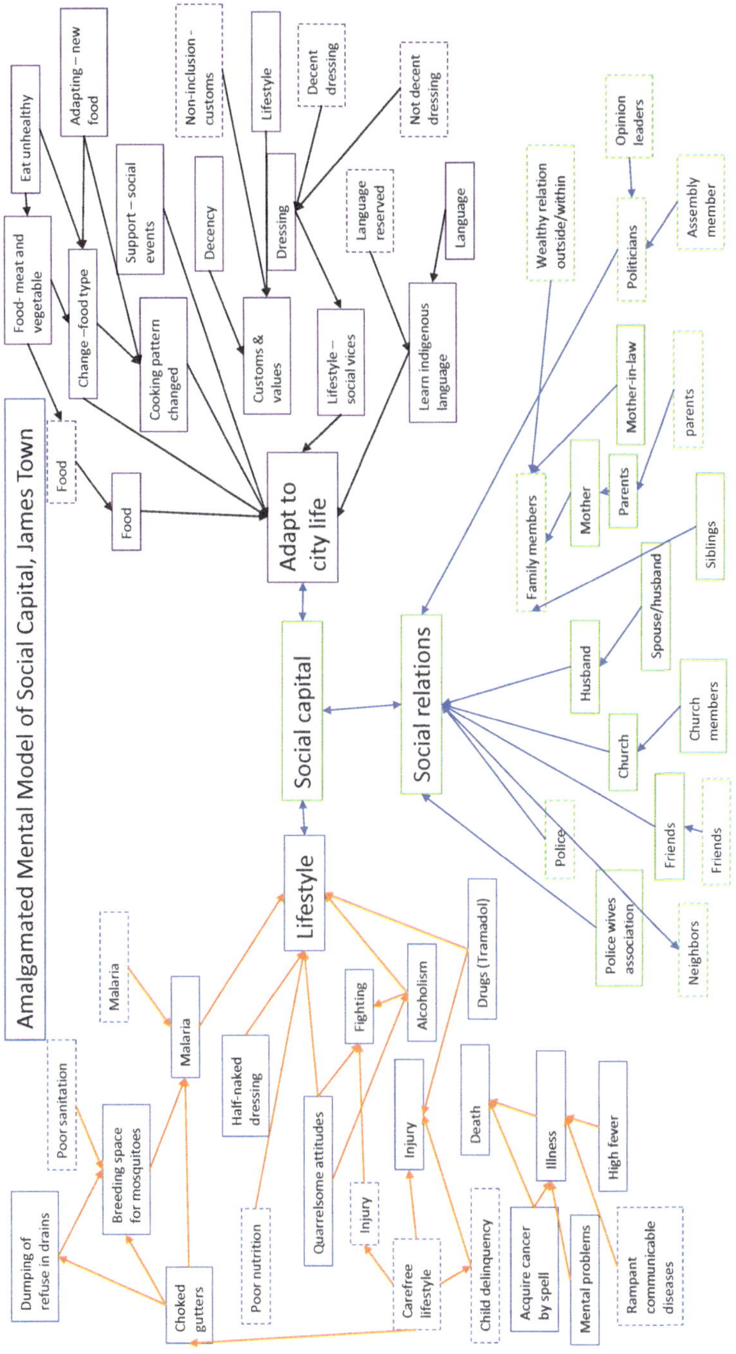

Fig. 7.1 Mental model of social capital, acculturation, lifestyle, and health, James Town

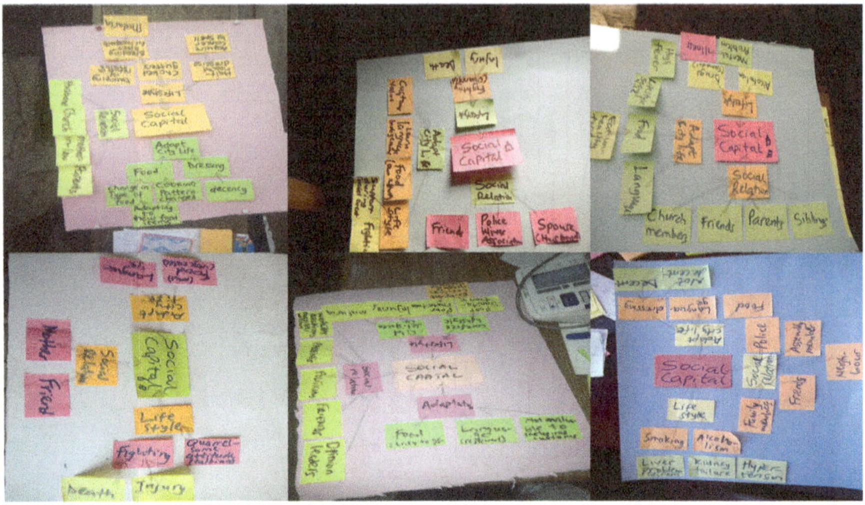

Fig. 7.2 Pictures of mental model of social capital, acculturation, lifestyle, and health, James Town

allowed him to diversify his food because from the village where he came, he "used to eat one type of food," the influence of social relations in liking certain kinds of unfamiliar food or trying new dishes or inculcating some foods into migrant regular diet seem very strong. A respondent who learned how to eat fufu (usually made of pounded cassava and plantain) when he migrated into the community said:

> My uncle in the community has also helped me a lot when it comes to food in the community. It was through my uncle that I was able to eat fufu that has been made from cassava which is not eaten back in my hometown. [Male; 36 years; 7 years in community]

In some cases, there was prestige associated with the food migrants have been introduced to by their social relations. Whereas the introduction to unfamiliar food by social relations seem to dominate, the attendant prestige associated with eating certain kinds of food was also important to migrants, especially if such food was not available to them at the place of origin. A couple of quotes from respondents illustrate such prestige as well as the role of social relations in enabling migrants to become comfortable eating food with which they were not previously familiar.

> I will say that by the help and advice of my friends and siblings here, I now eat white rice which I wasn't able to eat in my hometown because we did not have some there. [Male; 27 years; 6 years in community]

> I have also learned how to eat *tuozafi* [made from corn] through the help, influence, and encouragement of my friend when I came to this community. [Male; 43 years; 8 years in community]

Concerning eateries, a participant who asserted that he was familiar with all the foods in the city prior to migrating admitted to having to adjust to eat in local restaurants popularly called chop bars. He said:

> I had to learn and adjust to eat from chop bars when I arrived here; but I learnt how to eat most of the food I enjoy from this community back at my place before coming to settle here [Male; 55 years; 5 years in community]

In Old Fadama, work and job search seem to be crucial to adaptation to city life compared to our respondents in James Town. Social relations influence work search, work acquisition, as well as type of work. Migrants were introduced to work that their relations are involved, which might not require experience and/or formal training. The quotes below illustrate the influence of social relations in work acquisition and the gratitude of migrants for the opportunity. An untrained drug dispenser noted:

> I can tell [that] my brother has changed my life entirely for the better. Because when I came here he has employed me as a drug dispenser in his pharmacy shop. [Male; 23 years; 1 year in community]

Others are not privileged to obtain help immediately upon arrival. However, they form social relations that help to survive in the community and city at large. A respondent indicated that:

> I have learned to work hard since I came to live here. Though I did not receive any help or support from any of the mentioned relations initially when I came here, I had to join some friends to do different small small jobs [menial jobs] to get to where I am now. I really struggled to get myself a job later. [Male; 55 years; 5 years in community]

Learning a trade is helpful in ensuring occupational upward mobility in the slum. Trade like seamstress, masonry, and carpentry are common in the community. A migrant explains how a social relation was helpful in his skills acquisition as a tailor:

> About work, I will say that have been able to learn one new business [tailoring] from my friend which has helped me a lot financially. I am very grateful to him for that. Living in this community has made me to be job-conscious since everything done here seems to be in a rush. It is because of how busy the city life is like. [Male; 43 years; 8 years in community]

With respect to hygiene practices, health conscious social relations provide advice on the health hazards in the environment. A migrant recounted how his brother counseled him upon arrival about environmental risk to health in his immediate vicinity, which has helped him tremendously. He noted how he was surprised to find filth everywhere he came into the community.

Lifestyle and Health

Migrants mentioned lifestyle such as social unions, political involvement, peer pressure from too many friends, alcoholism, poor eating habits, smoking, congestion, hardwork ethics, and poor waste disposal practices. The interconnections among the lifestyles, as evidenced in the discussion of the migrants, are perceived to result in health outcomes—diseases and illnesses such as gonorrhea, HIV, diabetes, tuberculosis, waist pains, as well as injuries.

The responses on social unions yielded two diverging ideas. The discussion by migrants about social unions, as distinct from friends, referred to sexual relationships that exist in the community. Although some highlighted the negative consequences of sexual unions, others praised it. For participants who emphasized the adverse ramifications of such sexual relationships among the youth in the community, they mentioned implications of multiple concurrent and serial sexual unions for the spread of diseases such as HIV/AIDS and gonorrhea. However, some saw that not only as a way of keeping company but an unofficial marital union with the ability to ensure social stability and prevent the spread of sexually transmitted diseases. Such unions, usually referred to as consensual unions, are perceived as precursors for marriage ceremonies. A participant with a supporting view for the latter description of sexual relationships said:

> The community members also like… participating in marital-like sexual activities which I like so much; that he [the male counterpart of the union] [may] even want to carry out such marriage ceremony one day [in the future]. And to me, this union usually prevents one from engaging in promiscuous acts that can make him/her contract diseases like gonorrhea [Male; 36 years; 7 years in community]

Political engagement seems to have negative health outcome. A participant who was very particular describes in detail how the community is divided among political party lines. Such a division has resulted in several political unrests that degenerated into fighting, injuries, and death. Unfortunately, the political rift is along ethnic lines resulting in ethnic tensions already aggravated by ethnic enclaves and separation in the community. A migrant remarked:

> Community members argue so much on party political issues and this sometimes make them fight with each other. The fighting results in serious injuries. [Male; 36 years; 7 years in community]

Peer pressure and negative influences from too many friends are acknowledged as being partly responsible for crime and perceived criminal activity in the community. The presence of unofficial fraternity, distinct from ethnic group associations, is perceived as having the potential of leading to arrest and prosecution of innocent people in a group in the event of a crime, given the negative public perception about the community. A participant explained it as follows:

> Most people in the community like making too many friends and this is very bad because it can land innocent people in prison. This is because, if one person of the group does something bad and such person is wanted [by the police], you the friend who may not be part of such bad act of your friend could be added [as a suspect] due to the fact that you are always in the company of each other. [Male; 27 years; 6 years in community]

Risky health behaviors such as smoking, alcohol consumption, and frequent intake of sugary drinks were stated and linked to tuberculosis, diabetes, and heart diseases by migrants. Migrants decried the seemingly ubiquitous smoking of cigarettes and marijuana in the community and its ramifications for heart diseases. Some migrants indicated that nonparticipation in risky health behavior results in no illness, better lifestyle as well as peaceful life; they further demonstrated understanding of nuanced relationship between multiple risky health behavior, environment, and health outcomes. For example, with drinking and health, a participant remarked:

> The community members consume too much soft drinks and alcohol of which I think it is not good for their health because it can make them contract diabetes disease. Most community members especially the males smoke a lot, dress indecently, which I think is bad and needs not to be emulated. [Male; 44 years; 21 years in community]

The population density was perceived to be associated with health outcome. The high population density, according to a participant, is responsible for the rampant infectious diseases in the community. The pathways for the diseases include poor disposal of garbage and poor personal hygiene practices. A migrant indicated:

> Here, especially in this place, there are so many people. This area is very congested. I can tell you that back at home [in my hometown], it is not like that at all. People are not many like that throwing rubbish away by heart [at random]. And this makes us susceptible to airborne diseases such as TB [tuberculosis] because people are not neat. [Male; 55 years; 5 years]

Finally, connections were made between hardworking conditions in the community among migrants and negative health outcomes. Especially for physically demanding jobs, a participant who is a mason by occupation mentioned the stressful nature and physical energy exerted during work and linked that to waist pains suffered as a result. He mentioned how most migrants, both male and female, who are engaged occupations that require physical exertion of energy in the sun experience different forms of pains in their bodies. He said:

> I am now a hard-working person, and this has help me a lot. But working too hard will make you the individual get heart attack… hahaha! I tell you! I have waist pains like many here who do hard jobs. So many people have pains in the body too. [Male; 27 years; 6 years in community]

We depict the connections made by migrants on their individual conceptual maps into an amalgamated mental model. Please see Figs. 7.3 and 7.4.

Brief Deliberation on the Conceptual Mapping on Social Capital, Acculturation, Lifestyle, and Health for Old Fadama and James Town

Acculturation elements enabled by social relations include language acquisitions and comprehension, food, customs and values, as well as clothing styles. Varying forms of social capital are deduced from the social relations of migrants across the slums. Bonding social capital is evident, especially through nuclear family, whereas bridging social capital is apparent through friends who are either from the same ethnic city or different ethnicity including the indigenous population in the case of James Town. Linking social capital is manifest in connections to local political power brokers. Each of these forms of social capital influenced acculturation in different but sometimes similar ways. Bonding social capital mostly impacted food and economic acculturation while bridging social capital mostly influenced language acculturation, assimilation of customs and traditions, and learning of trade.

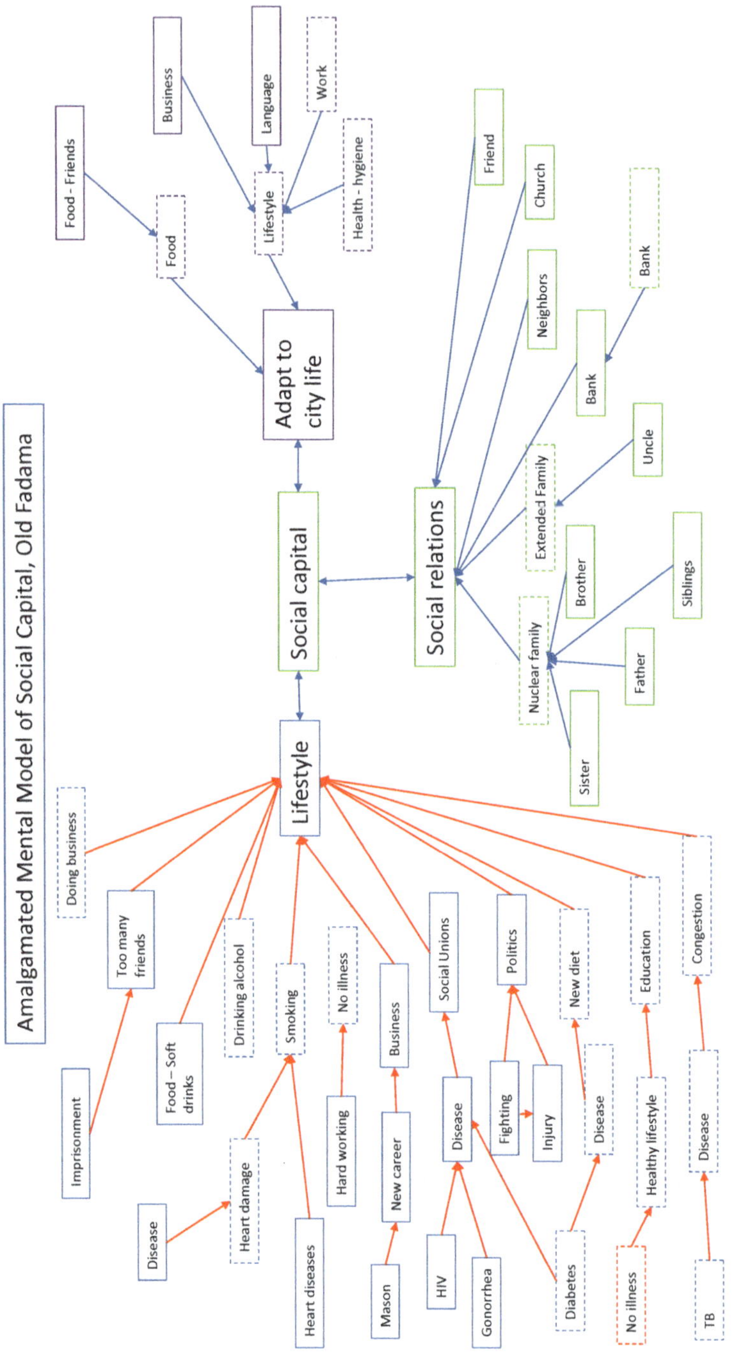

Fig. 7.3 Mental model of social capital, acculturation, lifestyle, and health, Old Fadama

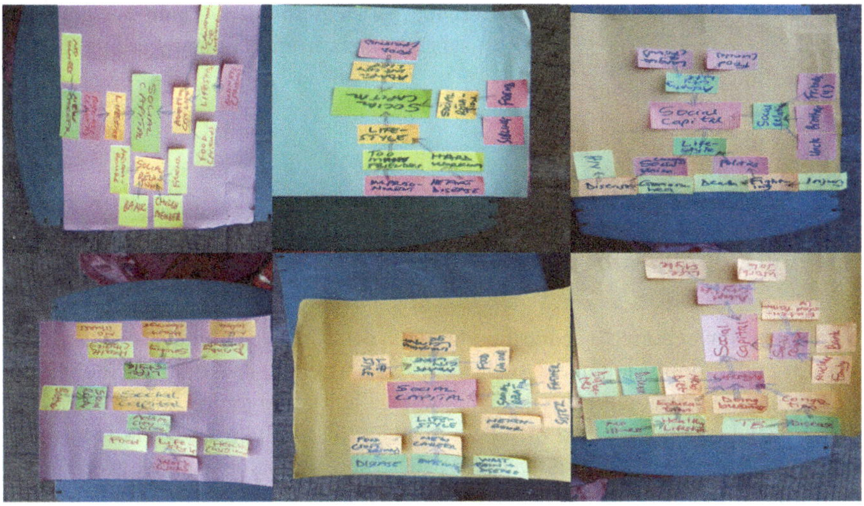

Fig. 7.4 Pictures of mental model of social capital, acculturation, lifestyle, and health, Old Fadama

Linking social capital, which was the least active form of social capital, manifested in advice and sense of political participation.

That notwithstanding, migrants seem selective with acculturation preferences. For example, while some love the food at the destination, they are not interested in learning the language. Overall, language was an important acculturation element although the kind of language perceived as crucial differed. Whereas Ga was the language of choice in James Town, obviously due to the place being the traditional home of the Ga people, in the more cosmopolitan poor Old Fadama, Twi was the preferred language. Most importantly, the place and the ability to function effectively as a member of the community influenced one's acculturation inclination regarding language.

Social relations have influenced migrants' lifestyles with both positive and negative ramifications for health. Such lifestyles include alcoholism, happy-go-lucky living, and poor eating habits; these were linked to diseases like hypertension, malaria and other communicable diseases, and sometimes death. As expected, the health professionals always made more sophisticated linkages and asserting the increasing patterns of diseases may be attributed to poor eating habits, poor nutrition, and substance abuse. A connection between lifestyle, spirituality, and health outcome was fashioned. Specifically, lack of decency was perceived as having the potential of invoking the wrath of deities with the consequences of inflicting culprits with cancer.

Finally, as has been discussed in preceding chapters, migrants usually link their environment to malaria burden. However, there seem to be a lack of understanding on malaria infection origins. The general poor sanitation and unhygienic surroundings are immediately linked to malaria whenever it comes up. While we did not probe to evaluate the nuanced pathways between the environment and malaria infections, the penchant to mention malaria whenever the poor environment conditions are mentioned may require further studies to separate facts from fictions regarding malaria transmission.

Results from Namuwongo Activities

Social Relations and Acculturation

Social relations reported as contributing to the needs of the migrants include neighbors, friends, family members such as sisters, brothers, and mothers, neighbors, church members, people from the same village before migration, and work mates. These social relations helped members to adapt to city life in different forms. The main forms of adapting that are aided by these social relations include work placement, provision of shelter, and introduction to other people in way of referring them for work.

Most of the respondents reported that they were from the same ethnic group and spoke the same language making it easy for them to adapt to the environment around them. They referred to is as a "natural connection" brought about by their commonality in language. Knowing the language of the people around helped them cope better, putting them on less pressure to learn the language spoken in central region-Luganda. One of the respondents noted that:

> When I came I didn't know what to expect since I was coming from Gulu. I had heard that people are eaten and killed in Kampala, but when I reached here I found people who spoke the same language as myself. That helped me settle down much more quickly and also find work very fast. [Male; 28 years; 8 years in community]

The ability to communicate in the same language also helped migrants settle and make friends much easily, which opened opportunities for them. Most indicated that finding such people enabled them to know what dangers they would face as one retorted:

> When I came from the village I had heard that all women who come to Kampala come to be prostitutes. I did not want to become a prostitute with painted nails and long hair. When I came and made some friends I realized that these are normal things that people in town do. They are not for prostitutes. [Female; 27 years; 5 years in community]

Some of the respondents indicated that their food consumption patterns have changed, though not so significantly because they cannot get their traditional delicacies since they came to Kampala as one retorted:

> The food I eat here is not the same as what we had back home but it is healthy. Here I have to improvise for most of the time because the money is not enough to buy what I want. What we used to pick and eat at home, we now have to buy. So I had to change and also learn eating bananas. But when I eat them in a few minutes am hungry again. Maybe my body will get used to. [Male; 32 years; 11 years in community]

As regards types of food cooked and their relationship to health, migrants reported that it is difficult for them to mind about healthy food because they cannot afford it when they cannot afford to have enough food.

> While am cooking and looking for food I also have to consider how nutritious the food it is, but sometimes it is not possible. But at least I get a lot of vegetables, more than I would get if am in the village. [Female; 35 years; 10 years in community]

As regards dressing, most respondents, especially females, reported that their way of dressing has changed and they have had to adapt to the ways of dressing they found in Kampala as one intimated:

> When I came to Kampala I knew that trousers were for prostitutes because in my village it is not allowed to wear trousers. But when I reached in Kampala I found most women wearing trousers and in most cases it was much safer because it is difficult for someone to sexually assault me. Because of that, even when am going to visit in the village I go home wearing trousers. The main thing is to make sure that am decent. [Female; 28 years; 8 years in community]

Lifestyles and Health

Lifestyle was explained as to how people conduct their lives in the community. Migrants reported listening to development ideas, drinking alcohol, multiple sexual partners, spreading of HIV/AIDS, peer group influence, prostitution, learning area language, freedom to do as one pleases, changing food habits, and negligence among others. The lifestyles mentioned here were reported to be intertwined and contributors to the diseases that were spread in the community for instance, HIV/AIDS, STIs, weight loss, and liver diseases as a result of consuming too much alcohol among others.

Migrants reported that being in the city allows them to live as individuals, choosing what to do when they want to do it. Males reported that unlike their villages, they are free to have many sexual partners that are not permanent and enjoy life on their own without the watchful eye of adults. They attributed the spread of HIV/AIDS and STIs such as gonorrhea and syphilis to this free lifestyle.

> Our men have more women than they could have in the village. Many women try to work and earn some money so the man is not worried about the home or the family since he knows that the woman will take care of them. Many men become irresponsible because of this lifestyle and choose to do as they please instead of being responsible. [Male; 38 years; 35 years in community]

> Women nowadays cannot be trusted; you can think you are the only man to be shocked. Women have more than one man, all of them providing for the same family. When one leaves another one comes in. Women are players. You will only know the problem when you find your genitals in pain and at the clinic they tell you that you have gonorrhea. [Male; 25 years; 4 years in community]

Migrants also reported that the free lifestyle has led many women into prostitution and yet back in their homes they were decent girls.

> My sisters from Nebbi did not know anything about prostitution but now they are experts all working in Kabalagala area as super prostitutes. And sometimes when you find them you cannot even recognize them because they have bleached their skin and they wear long hair extensions [Male; 29 years; 5 years in community]

Migrants reported that such free lifestyle leads to family breakdown and the girls involved do not want to get married, because they would not be able to earn. The children they give birth to also end up becoming irresponsible and taking drugs.

Migrants reported that in some cases too much alcohol consumption was attributed to ill health and delinquency among the community members. They reported that in homes where there is a lot of alcohol consumption, there is a lot of domestic violence and couples end up separating or divorcing.

> In most of these homes where both partners consume alcohol, there are constant fights. Both partners will go to the bar and drink but when they come back, they start fighting. Sometimes they beat up all their children and the whole household is in chaos. This causes children to run away from home onto the streets. [Male; 33 years; 10 years in community]

> In this community, women drink more than men. They go to the bars in the morning and leave late at night having fought everyone in the bar. They forget that they have children, families and other things to do. Their children can run around the whole community naked and fall in trenches only to be picked by strangers. Their children are always in the clinic and their mothers do not care. When the children turn to drugs and start prostitution, then there is no one to blame. [Female; 35 years; 10 years in community]

The health professionals also interlinked the alcohol consumption with drug abuse, multiple sexual partners, and poor feeding habits. They reported that some of the people consume alcohol on empty stomachs, abuse drugs, and will have sexual partners in all kinds of places hence spreading diseases.

> We are seeing many cases of hypertension among people who are busy consuming alcohol and abusing drugs. So even if someone is on drugs for hypertension, they cannot get well because they are not feeding well or appropriately. Even if you talk, they will not listen, and some have ended up losing their lives because of this, resulting into liver and kidney diseases. [Female; 40 years; 5 years in community]

> A person will sit with friends on five to ten kilos of pork and eat accompanied by waragi (Ugandan gin). They will do this every day for seven days a week nonstop. Meanwhile at home he will not have left food for the family but goes back drunk and satisfied. Because of the situation at home, friends of the wife will advise on how she can make quick money. Before you know it, the man is bed ridden because of pressure and the woman is sick of STIs if she is lucky. They only can afford cassava chips and dry tea. Definitely they will die. [Female; 38 years; 15 years in community]

This shows a strong linkage between health and lifestyle among the study group but also points to the dangers that accrue from the changes in the lives of the migrants.

The connections made by migrants on their individual conceptual maps are organized into an amalgamated mental model (Fig. 7.5).

Fig. 7.5 Mental models on social capital, lifestyle, and health, Namuwongo

Results from Kinawataka Activities

Social Relations

In Kinawataka, social relations were considered as a reason for someone settling in one part of the slum instead of another. Among the social relations that were considered as important to the migrants were family members such as brothers, sisters, and friends, tribemates, local council officials, business partners, business development groups, self-help groups, and workmates. These social relations were considered important for the contribution they made to acclimatization with city life for the migrants. Migrants reported that social relations helped them access food, accommodated them, gave them advice, giving information about the environment, reference to work places, and provided employment.

Migrants reported that for relatives, one has a natural bond with them and they will help because of the blood relationship. In Kinawataka, migrants indicated that they tend to come from different tribes but the same ethnic group, which makes communication, culture, and customs easy to adapt to. They reported that the commonly spoken language is Lugishu, which all ethnic groups understood. Coming from the same tribe and being in one area enabled them settle quickly and find work in the neighboring communities

> When I came I was scared that I would not have food to eat because I didn't know anyone. But when I reached here and everyone was speaking a familiar language I found that they were my tribes mates so it was easy to find a house and a job to settle down. [Female; 27 years; 5 years in community]

Coming from the same ethnic group enabled more interaction between migrants in that they identified with those who spoke the same language, ate the same food, and did similar cultural things.

> I first thought that I would not find people who spoke my language or who understood or knew where I come from. The friend I came with advised that we come to this area because there were many people from the East. When I got into this community it was like being in the village back home, I was able to communicate freely and settle in the community. I set up a fruits vending business and it has sustained me. [Female; 35 years; 10 years in community]

The food and alcohol consumption patterns were part of the activities that brought the groups together because they meet often and drink together as one retorted:

> Most of us from the East and North like to drink 'malwa'(locally fermented alcoholic drink). When we are drinking we discuss different issues and also agree on some community programs. Some people call it drinking alcohol but for me it is a social gathering. When I came the first time, I made my first friends at the malwa drinking joint and that has enabled many other relationships to emerge from it. Malwa is part of our culture and it brings us together as community members. [Male; 34 years; 8 years in community]

Migrants reported that the food they consume has been altered because of the situation in the city but when one travels to the village and brings they share the little that has come from there. They indicated that at some of the bars, they also serve local delicacies but at home the food was adapted to their economic circumstances.

> In Kampala, all food is available from all kinds of place, it is one's ability to afford it that makes the difference. For instance, in the market where I work, you can find 'malewa' (bamboo shoots that are a delicacy in Eastern Uganda). Back home, these grow wild and you pick as you please, but here they are in the market. So instead I eat 'posho' (corn meal) and beans like a prisoner because that's what I can afford. [Male; 36 years; 10 years in community]

> When the man comes back home, he simply wants food and he knows that the situation has changed from the village. What you serve is what he will eat because he knows that all of you cannot afford the food that is consumed in the village. If you are lucky and he passes by the bar, he eats and comes back satisfied. That way he avoids the food at home. Even if we know that vegetables and fruits are important because the nurse tells us all the time, it is cheaper to buy soda than milk. [Female; 30 years; 10 years in community]

Most of the things that they did back at their homes were reportedly changed with their moving to the city because the situations are different. For instance, as regards dressing one member intimated that:

> Our ways of dressing has changed especially the women. Women here are almost naked, they have no clothes. They are copying modern dressing but in the end are naked. When you come in the evening at the bar, you find many girls just waiting for customers because they

> work as sex workers. If you dare tell them anything they will abuse you as if they have never seen you and you keep quiet. They even gang up and smear your house with feaces so that flies can come all over your house and the smell is unbearable too. Such things could not be seen in our village. [Male; 30 years; 6 years in community].

> When I came here I wanted to find a job and work and make a decent living. I stayed with a lady who had been kind enough to give me a place to stay. After looking for a job for a long time without success I had to use my head. My house mate used to come back with some money every evening, so I asked her what she does, and she introduced me to sex work. So, what we wear is like uniform, the clothes have to be revealing so that customers can see that you are working. If they don't then you will sleep hungry. [Female; 24 years; 4 years in community]

The dressing was dictated by the need to survive, the nature of one's work, and the need to attract customers, as opposed to decency.

Lifestyles and Health

Migrants in Kinawataka reported that lifestyle included things that defined their ways of life and how they conducted themselves. They reported that drug abuse, poor waste disposal, quarrels, fighting, spread of diseases such as HIV/AIDS, revenge, and theft as some of their livelihood patterns.

Migrants reported that the city life has exposed them to diseases that they would not have otherwise contracted. They explained that in the city everyone doesn't want to disturb the peace of another person because they might go away and leave one to pay bills alone.

> In Kampala if I complain about the man having many women, he will have many waiting for him and I will have lost a portion of his income. Even if am aware that he has diseases, I will treat them as they come and remain with him. Sometimes you are lucky, and you don't get sick. So, we stay together because we cannot tamper with each other's peace and keep contracting diseases. [Female; 27 years; 4 years in community]

> If a man tells me that I cannot do my work, then I will leave and find another one who accepts. Both of us can see the situation and understand it. We know how the diseases are contracted, but we are also taught how to prevent them. So, we prevent them and protect ourselves but do what we can to survive. [Female; 24 years; 5 years in community]

Such lifestyles were said to have increased the rate of drug and alcohol abuse in the area and having children that no one wants to care for.

> When you walk in the dark corridors after 7 pm, you will be shocked at what you find. People smoking all kinds of things, some of which you cannot even understand. From there they just sleep recklessly as if they are getting medals and father children with every other woman without a sense of responsibility. The child affairs office of the Local Council is a very busy one because of such people. Sometimes these children get sick of diseases like diarrhea and they die because both mother and father are too high to care for the babies. [Female, Nurse; 35 years; 6 years in community]

In some cases, migrants reported that some community members have been hospitalized because of drug abuse. They reported that they smoke drugs until they become mentally ill and cannot live in the community any longer. They either roam the community as mentally ill people, or they are taken to the mental hospital by relatives and friends.

Some reported that there is increased violence in the community because of some of these lifestyle choices:

> In Katoogo, one has to be careful what they say about another. Youths who consume drugs gang up and molest you if they hear that you said anything about them. The Local council has tried to get them arrested but before you know it, they are back in the community roaming the whole area and threatening violence. So, to keep peace, you keep quiet and mind your own business. [Female; 30 years; 5 years in community]

> Some men reach the bars at 7 am in the morning and drink the entire day. Most of them cannot afford to buy food so they console themselves with drinking. At the end of the day someone is too drunk to walk. If they are lucky they manage to reach home. Doing this continuously leads to multiple problems and many times they have to be hospitalized with complications in their stomach. (Male, 40 years, 20 years in community).

The linkages between lifestyle and health were very apparent in Kinawataka, and the connections between the different conceptual maps have been organized into an interlinked mental model (Fig. 7.6).

Fig. 7.6 Mental models on social capital, lifestyles, and health in Kinawataka

References

Berry JW, Phinney JS, Sam DL et al (2006) Immigrant youth: acculturation, identity, and adaptation. Appl Psychol 55(3):303–332

Braun M, Glöckner-Rist A (2012) Patterns of social integration of Western European migrants. J Int Migr Integr 13(4):403–422

Djamba YK, Kimuna SR (2012) The labor force participation and earnings gap among African immigrant women in the USA. J Int Migr Integr 13(4):481–501

Frank R, Heuveline P (2005) A crossover in Mexican and Mexican-American fertility rates: evidence and explanations for an emerging paradox. Demogr Res 12(4):77

Hawkins RL, Maurer K (2010) Bonding, bridging and linking: how social capital operated in New Orleans following Hurricane Katrina. Br J Soc Work 40(6):1777–1793

Kalmijn M, Van Tubergen F (2006) Ethnic intermarriage in the Netherlands: confirmations and refutations of accepted insights. Eur J Population 22(4):371–397

Mondain N, Lardoux S (2013) Transitions to adulthood among first generation sub-Saharan African immigrant adolescents in Canada: evidence from a qualitative study in Montreal. J Int Migr Integr 14(2):307–326

Nakhaie MR, Kazemipur A (2013) Social capital, employment and occupational status of the new immigrants in Canada. J Int Migr Integr 14(3):419–437

Raza M, Beaujot R, Woldemicael G (2013) Social capital and economic integration of visible minority immigrants in Canada. J Int Migr Integr 14(2):263–285

Sanders J, Nee V, Sernau S (2002) Asian immigrants' reliance on social ties in a multiethnic labor market. Soc Forces 81(1):281–314

Teja Z, Schonert-Reichl KA (2013) Peer relations of Chinese adolescent newcomers: relations of peer group integration and friendship quality to psychological and school adjustment. J Int Migr Integr 14(3):535–556

Chapter 8
Social Capital and Health

Abstract This chapter advances the discourse on migration, social capital, and health. It explores the association between social capital and health literacy in two urban slums in the city of Accra (Old Fadama and James Town). Conceptualizing social capital of migrants as resources derived from social relations that enable them to withstand stressors, we examine the relative effects of social capital on health literacy on infectious and noncommunicable diseases. Regarding resources, we focused on materiality (e.g., food, shelter, money), information for successful living, and emotional support (empathy, genuine concern, compassion). A distinction was made between resources accruing in general and resources accumulating in times of illness. A survey was used in eliciting data from a random sample of 150 individuals per slum. Methods of bivariate analyses used are Pearson correlation and analyses of variance (ANOVA). To understand the relative effects of social capital on health literacy, multiple regression was performed. We found that while increase in social capital was linked with increase in total health literacy in Old Fadama and health literacy on hypertension in James Town, increase in social capital was linked with decrease in health literacy on infectious diseases in both James Town and Old Fadama.

Keywords Health literacy · Disease knowledge · Social capital · Materiality · Infectious disease · Hypertension · Ghana · Social relations · Health information · Emotional support

Introduction

In this chapter, we demonstrate the association between social capital and health with examples from two different places thereby illustrating the importance of place in the social capital and health nexus discourse. This chapter establishes how social networks of internal migrants provide a buttress to the shortfalls of broken social

R. A. Tutu, J. D. Busingye, *Migration, Social Capital, and Health*,
Global Perspectives on Health Geography,
https://doi.org/10.1007/978-3-030-24693-8_8

and relational networks after leaving their place of origin. It shows the relevance and nature of social networks to internal migrants given the relational gaps that are left by disentangled networks from their areas of origin.

As discussed in Chap. 2, evidence of the relationship between social capital and health among migrants is ubiquitous. Studies on the connections between social capital and health provide avenues for interrogating ways by which health may be improved (Ronconi et al. 2012), especially given evidence of the mixed relationship between social capital and health (Kunitz 2004). For instance, while social capital has been linked with positive health (Iversen 2008; Petrou and Kupek 2008), it is inversely related to certain mental health outcomes including psychological distress (Scheffler et al. 2007). In an exploratory study that examined the association between social capital and self-reported health, Iversen (2008) found that community scale social capital in the form of voting participation in local elections was positively associated with self-rated health although sport associations was negatively related with self-assessed health status. More so, when social classes are controlled for, the social capital and health relationship show more dynamism. Verhaeghe et al. (2012) found that due to the differences in resource access among different social classes, while resources derived from friends and family among the higher social classes impacted self-rated health positively, resources from social relations among the working class seemed to impact self-rated health adversely. In Scheffler et al.'s (2007) study, area scale social capital was adversely associated with individual scale psychological distress among people whose family income is lower than the median income. These findings buttress Kunitz's (2004) assertion that context is crucial in understanding the social capital and health nexus. Whereas social capital at the individual scales has been demonstrated to have a strong positive effect on health (Mohnen et al. 2012), social capital at different contextual scales is mixed. That notwithstanding, higher social capital at the neighborhood scale is associated with better health (Mohnen et al. 2011; Diez Roux and Mair 2010). Neighborhood social capital is beneficial, especially in the urban context in many ways (Mohnen et al. 2011). Specifically, social bonds between neighbors comprising social capital, which is embodied in stronger social cohesion and reciprocity between neighbors, are protective factors against depression (Mair et al. 2008; Gary et al. 2007; Latkin and Curry 2003).

Consequently, although resources obtained from social networks are important for health outcome, the flow of actual resources from social relations and the specific social relations that help buffer specific health stressors are not fully understood across slums. Therefore, for this chapter, we examine the relationship between social capital and health literacy on infectious and noncommunicable diseases.

Social Relations and Social Capital Constructs

On social relations or networks from which social capital is acquired, based on our previous and current field research knowledge of the communities and literature, we asked questions about friends; family (nuclear)—brothers, sisters, father, mother, and step-parents; relatives—uncles, aunties, and cousins; church—including church members and church leaders; ethnic association—including group members, group leaders, and group sympathizers; as well as social unions—boyfriends, girlfriends, sugar daddies, and sugar mummies.

Concerning resources derived from the social networks, we focused on materiality, information, and emotional support in general terms, as well as when migrants were sick. Specifically, in general terms, we asked (1) whether migrants receive material help (e.g., money, clothing, food, shelter) from any of the categories of people (social networks listed previously) in their community; (2) whether migrants receive information about how to live successfully in their community from any of their social networks in their community; and (3) whether migrants receive emotional support (empathy, genuine concern, compassion) from any of their social networks who are in their community. We then asked separately for when they fall sick. Precisely, we asked (1) whether migrants receive material help (e.g., money, clothing, food, shelter) from their social relations who live in their community when they get sick; (2) whether migrants receive information about how to avoid being sick in their community from any of their social networks in their community; and (3) whether migrants, when sick, receive emotional support (empathy, genuine concern, compassion) from any of their social networks who are in their community. Responses were either yes—1 or no—2 for each of the categories of social relations. We hypothesize that migrants with high social capital are more likely to have higher scores on health literacy.

Analyses

We created total scores for "social capital general" and "social capital when sick." This was the statistical sum of the scores per each social relation item measuring materiality, information, and emotional support. Each specific social relation item is assumed to have the same contributory weighting to the overall score. Note that in view of the coding described above, higher score connotes poor social capital.

Subsequently, Pearson correlations were run to assess the association between each of the social capital scores (i.e., social capital general scores and social capital when sick scores) and total health literacy scores, health literacy scores on infectious diseases, as well as health literacy scores on noncommunicable diseases. While ANOVA was performed to examine relationship between the social capital scores and discrete sociodemographic variables—education, religious affiliation, marital status, occupation, and ethnicity—Pearson correlation was used to explore the association between the social capital scores and continuous demographic variables—age, age at migration, duration of stay in the community, and household size.

To understand the relative effects of social capital on health literacy, multiple regression was performed. This was dependent on the associations that were established after the initial screening using Pearson correlation and ANOVA. Dummy variables were created for the demographic predictors that needed to be controlled for in the multiple regression analyses.

Table 8.1 Correlation and ANOVA of health literacy, social capital, and demographics

	Total health literacy	Total health literacy—IFD	Total health literacy—NCD
Social capital (general)	−0.149	0.15	−0.126
Social capital (when sick)	−0.065	0.193*	−0.177*
Age	−0.147	−0.188*	0.228*
Sex	1.445	1.182	5.454*
Education	4.909**	8.033**	6.440
Marital status	3.105*	5.118***	1.267
Occupation	4.931***	3.447*	1.565

*$p < 0.05$, **$p < 0.001$, ***$p < 0.01$

Results from James Town Study

Bivariate Association Between Health Literacy, Social Capital, and Demographic Characteristics of Migrants

For the James Town data, Pearson correlation results showed no association between social capital (general) and health literacy (Table 8.1). However, social capital (when sick) was found to be significantly associated with total health literacy on infectious diseases and total health literacy on noncommunicable diseases. Age of the migrants was also significantly associated with total health literacy on infectious diseases and total health literacy on noncommunicable diseases. Results from the ANOVA show significant association between sex, education, and total health literacy on noncommunicable diseases and between education, marital status, occupation, and total health literacy on infectious diseases.

Multiple Regression: Health Literacy, Social Capital, and Demographic Characteristics of Migrants

With the focus of this chapter being to examine the relative effects of social capital on health literacy, multiple regression models were run. The first series of models had health literacy on infectious disease (gonorrhea) as dependent variable and "social capital when sick" as main independent variable controlling for significant confounding factors. The second series of models had health literacy on noncommunicable diseases as outcome variable and "social capital when sick" as the predictor variable controlling for significant demographic variables.

The results of the regression examining whether "social capital when sick" predicts health literacy on infectious diseases (Table 8.2; model 2) indicated that the model explained 6.5% of the variance and that the model was a significant predictor of health literacy on infectious diseases ($F = 5.133$, $p = 0.007$). Controlling for age,

Table 8.2 Multiple regression analyses for variable predicting health literacy on infectious disease, James Town

	Model 1			Model 2			Model 3			Model 4		
	B	Std. error	Beta	*B*	Std. error	Beta	*B*	Std. error	Beta	*B*	Std. error	Beta
Social capital (when sick)	0.384*	0.16	0.19	0.346*	0.159	0.174	0.248	0.153	0.125	0.237	0.153	0.119
Age				−0.06*	0.029	−0.168	−0.017	0.028	−0.047	0.009	0.033	0.025
Education												
Primary education							2.143	1.735	0.127	2.376	1.74	0.141
Junior high school							1.035	1.364	0.099	0.642	1.394	0.062
Senior high school							4.762**	1.528	0.367	4.117*	1.632	0.317
Tertiary							5.734**	1.688	0.383	4.044*	1.93	0.27
Marital status												
Cohabiting										−2.039	1.643	−0.107
Separated/divorced/widowed										−3.612*	1.503	−0.246
Married										−0.921	1.051	−0.089
Occupation												
Sales										−2.634	1.552	−0.254
Manual										−2.344	1.717	−0.163
Unemployed/student										−2.496	1.509	−0.213

$*p < 0.05$, $**p < 0.01$

Table 8.3 Multiple regression analyses for variable predicting health literacy on noncommunicable disease, James Town

	Model 1			Model 2		
	B	Std. error	Beta	*B*	Std. error	Beta
Social capital (when sick)	−0.189*	0.087	−0.2	−0.182*	0.084	−0.17
Age				0.043**	0.015	0.225
Sex				−1.646**	0.586	−0.219

*$p < 0.05$, **$p < 0.01$

"social capital when sick" contributed significantly to the model ($B = 0.346$, $p = 0.038$). That is, there is a 0.346 increase in total health literacy score on infectious disease for each unit increase in poor social capital. For each additional unit increase in age, total health literacy on infectious disease decreases by 0.060.

In model 3 (Table 8.2), we controlled for age and education. The model explained 19.9% of the variance in the outcome variable, and the model was a significant predictor of health literacy on infectious diseases ($F = 5.916$, $p < 0.001$). While senior high school education ($B = 4.762$, $p = 0.002$) and tertiary school education ($B = 5.734$, $p = 0.001$) contributed significantly to the model, "social capital when sick," age, primary education, and junior high school education did not. Migrants who are high school educated have mean health literacy on infectious disease score that is 4.762 higher than those with no education. Respondents who are tertiary school educated have mean health literacy on infectious disease score that is 5.734 higher than migrants with no education.

Model 4 (Table 8.2) added all other dummy variables of significant demographic variables—marital status and occupation. The model explained 24.8% of the total variance in the response variable. The model was a significant predictor of the outcome variable ($F = 3.765$, $p < 0.001$). Although "social capital when sick," age, being primary school educated, having junior high school education, cohabiting in a union, being married, and occupation lose their predictive powers, being senior high school educated ($B = 4.117$, $p = 0.013$), having tertiary school education ($B = 4.044$, $p = 0.038$), and being divorce/separated ($B = -3.612$, $p = 0.018$) contributed significantly to predicting health literacy on infectious diseases. Migrants with senior high school education have mean health literacy on infectious disease score that is 4.117 higher than those with no education, while migrants who are tertiary school educated have mean health literacy score that is 4.044 higher than migrants with no education. Respondents who were divorced or separated have mean health literacy on infectious disease score that is 3.612 lower than migrants who are single.

The second set of models had health literacy on noncommunicable diseases as the response variable. In all, two models were run given the results of the bivariate analyses. The results of the regression examining whether "social capital when sick" predicts health literacy on noncommunicable disease—hypertension (Table 8.3; model 2) indicated that the model explained 12.3% of the variance and that the model was a significant predictor of health literacy on noncommunicable disease ($F = 6.795$, $p < 0.001$). Controlling for age and sex, "social capital when sick" contributed significantly to the model ($B = -0.182$, $p = 0.038$). That is, there is a 0.182

decrease in total health literacy score on hypertension for each unit increase in poor social capital. Age ($F = 0.043$, $p = 0.005$) and sex ($F = -1.646$, $p = 0.006$) also contributed significantly to the model. For each additional unit increase in age, total health literacy on hypertension increases by 0.043. Males had an estimated mean of health literacy on hypertension that is 1.646 lower than that of females.

Results from Old Fadama Study

Bivariate Association Between Health Literacy, Social Capital, and Demographic Characteristics of Migrants

For Old Fadama, Pearson correlation results showed association between "social capital general" and total health literacy as well as health literacy on infectious disease—upper respiratory infection (common cold) (Table 8.4). However, social capital was not found to be significantly associated with total health literacy on noncommunicable disease—hypertension. Migrants' household size was significantly associated with total health literacy but not significantly associated with health literacy on infectious diseases and total health literacy on noncommunicable diseases. Results from ANOVA show significant association between education and total health literacy on infectious disease and between religiosity and total health literacy on infectious disease.

Multiple Regression: Health Literacy, Social Capital, and Demographic Characteristics of Migrants

To assess the relative effects of social capital on health literacy, multiple regression models were run. The first series of models had total health literacy as dependent variable and the social capital variables ("social capital general" and "social capital when sick") as main independent variable controlling for significant confounding

Table 8.4 Correlation and ANOVA of health literacy, social capital, and demographics, Old Fadama

	Total health literacy	Total health literacy—IFD	Total health literacy—NCD
Social capital (general)	−0.189*	0.209*	−0.156
Social capital (when sick)	−0.224**	0.211**	−0.140
Household size	0.189*	0.158	0.048
Education	9.243***	4.332***	1.514
Religiosity	0.377	2.424*	2.033

*$p < 0.05$, **$p < 0.01$, ***$p < 0.001$

Table 8.5 Multiple regression analyses for social capital (general) and total health literacy, Old Fadama

	Model 1			Model 2		
	B	Std. error	Beta	*B*	Std. error	Beta
Social capital (general)	−0.626*	0.267	−0.189	−0.72*	0.243	−0.218
Household size				0.529*	0.201	0.188
Education						
Primary education				1.695	1.333	0.103
Junior high school				1.772	1.157	0.13
Senior high school and higher				8.292***	1.362	0.492

*$p < 0.05$, ***$p < 0.001$

factors in Table 8.5. The second series of models had health literacy on infectious diseases as response variable and the social capital variables as the predictors controlling for significant demographic variables, per Table 8.4.

The results of the regression examining whether "social capital general" predicts total health literacy (Table 8.5; model 1) indicated that the model explained 3.6% of the variance and that the model was a significant predictor of total health literacy ($F = 5.494, p = 0.020$). Adjusting for confounding factor (household size and education), in model 2, "social capital general" contributed significantly to the model ($B = -0.720$, $p = 0.004$). That is, there is a 0.720 decrease in total health literacy score for each unit increase in poor social capital.

Household size and education were also significant predictors of total health literacy score (Table 8.5, model 2). The model explained 24.6% of the variance in the outcome. Regarding household size, there was a 0.529 increase in total health literacy score for each unit increase in household size. While primary education and junior high education did not contribute significantly to the model, senior high school or higher education contributed significantly to the model. Migrants who are high school or higher educated have mean total health literacy score that is 8.292 higher than those with no education.

In the series of models assessing whether "social capital when sick" predicts total health literacy, the first model explained 4.4% of the total variance. The model was a significant predictor of the response variable ($F = 7.808$, $p = 0.006$). Model 2 controlled for household size and education. This model explained 27.3% of the total variable and was a significant predictor of total health literacy ($F = 12.181$, $p < 0.001$). All the independent variables were significant predictors of total health literacy. For each additional increase in poor social capital (when sick), total health literacy reduced by 0.960. Concerning household size, there is a 0.505 increase in total health literacy score for each unit increase in household size. Migrants who have senior high school education and higher have mean total health literacy score that is 8.658 higher than migrants with no education. In model 3, both social capital variables were controlled for in addition to the demographic predictors. The model explained 26.8% of the total variance. The model fitted well ($F = 10.083, p < 0.001$). While "social capital when sick," household size, and education maintained their predictive powers, "social capital general" was not statistically significant.

Next, multiple regression models were run to investigate whether "social capital general" and "social capital when sick" could significantly predict health literacy on infectious disease—upper respiratory infection (common cold). The first set of models included "social capital general" as the main predictor controlling for demographics that were significantly associated with health literacy on infectious disease at the bivariate stage. Model 2 (Table 8.6) shows that "social capital general," education, and religiosity are significant predictors of health literacy on infectious disease. The model explained 17.3% of the variance and that the model was a significant predictor of health literacy on infectious disease ($F = 3.699$, $p = 0.001$). Migrants with senior high school education or higher have mean total health literacy score that is 2.170 higher than migrants with no education. Respondents who always practiced the tenets of their faith have mean total health literacy score that is 4.141 higher than migrants who reported "not at all" to practice the tenets of their faith.

Two models (Table 8.7) were run to assess whether "social capital when sick" could significantly predict health literacy on infectious disease. "Social capital when sick" maintained its predictive capacity in both models. While "social capital when sick" and religiosity contributed significantly to the model, education did not. Model 2, which controlled for the demographics, explained 17.6% of the variance and that it was a significant predictor of health literacy on infectious disease ($F = 3.755$, $p = 0.001$).

There is a 0.516 increase in total health literacy score on infectious disease—upper respiratory infection—for each unit increase in poor social capital, whereas migrants who always practice the tenets of their faith have mean total health literacy score on infectious disease that is 4.345 higher than migrants who reported "not at all" to practice the creeds of their faith.

Brief Deliberation on Social Capital and Health Literacy

Social relations have served as important resource for migrants in diverse ways. For this chapter, we explored the contribution of social relations in providing material resources (e.g., money, food, shelter, clothing), emotional support (e.g., comfort, empathy, genuine concern, compassion), and information on living successfully both in general and when migrants were sick. Subsequently, social capital indexes were created and association with migrant's health literacy was assessed.

While social capital was significantly associated with health literacy at the bivariate stage, it did not always predict health literacy. The association between social capital and health literacy was not always positive. While increase in social capital was linked with increase in total health literacy in Old Fadama and health literacy on hypertension in James Town, increase in social capital was linked with decrease in health literacy on infectious diseases in both James Town and Old Fadama. Future studies will need to define clear pathways by which each social capital construct may be impacting health literacy while controlling for demographic and context variables.

Table 8.6 Multiple regression analyses for social capital (general and when sick) and total health literacy, Old Fadama

	Model 1			Model 2			Model 3		
	B	Std. error	Beta	*B*	Std. error	Beta	*B*	Std. error	Beta
Social capital (general)							0.049	0.413	0.015
Social capital (when sick)	**−0.794*****	0.284	−0.224	**−0.96*****	0.253	−0.271	**−1.003*****	0.438	−0.283
Household size				0.505*	0.197	0.18	0.503*	0.199	0.179
Education									
Primary Education				1.998	1.305	0.122	2.019	1.321	0.123
Junior high school				1.934	1.137	0.142	1.927	1.142	0.142
Senior high school/higher				8.658***	1.345	0.513	8.668***	1.352	0.514

$*p < 0.05$, $***p < 0.001$

Table 8.7 Multiple regression analyses for social capital and health literacy on infectious disease, Old Fadama

	"Social capital general" and infectious disease						"Social capital when sick" and infectious disease					
	Model 1			Model2			Model 1			Model 2		
	B	Std. error	Beta	*B*	Std. error	Beta	*B*	Std. error	Beta	*B*	Std. error	Beta
Social capital (general)	0.483*	0.186	0.209	0.472*	0.193	0.204						
Social capital (when sick)							0.521*	0.199	0.211	0.516*	0.204	0.208
Education												
Primary Education				−0.813	1.017	−0.071				−1.011	1.01	−0.088
Junior high school				0.817	0.878	0.086				0.785	0.878	0.082
Senior high school/higher				2.17*	1.041	0.184				2.006	1.046	0.17
Religiosity												
Less often				0.422	2.273	0.021				0.791	2.267	0.04
Often				1.687	1.952	0.143				1.879	1.931	0.159
Very often				2.39	1.818	0.261				2.566	1.81	0.28
Always				4.141	1.813	0.442				4.345*	1.807	0.463

$*p < 0.05$, $***p < 0.001$

References

Diez Roux AV, Mair C (2010) Neighborhoods and health. Ann N Y Acad Sci 1186(1):125–145

Gary TL, Stark SA, LaVeist TA (2007) Neighborhood characteristics and mental health among African Americans and whites living in a racially integrated urban community. Health Place 13:569–575

Iversen T (2008) An exploratory study of associations between social capital and self-assessed health in Norway. Health Econ Policy Law 3(Pt 4):349–364

Kunitz SJ (2004) Social capital and health. Br Med Bull 69(1):61–73

Latkin CA, Curry AD (2003) Stressful neighborhoods and depression: a prospective study of the impact of neighborhood disorder. J Health Soc Behav 44:34

Mair C, Roux AD, Galea S (2008) Are neighbourhood characteristics associated with depressive symptoms? A review of evidence. J Epidemiol Community Health 62(11):940–946

Mohnen SM, Groenewegen PP, Völker B, Flap H (2011) Neighborhood social capital and individual health. Soc Sci Med 72(5):660–667

Mohnen SM, Völker B, Flap H, Subramanian SV, Groenewegen PP (2012) You have to be there to enjoy it? Neighbourhood social capital and health. Eur J Public Health 23(1):33–39

Petrou S, Kupek E (2008) Social capital and its relationship with measures of health status: evidence from the Health Survey for England 2003. Health Econ 17:127–143

Ronconi L, Brown TT, Scheffler RM (2012) Social capital and self-rated health in Argentina. Health Econ 21(2):201–208

Scheffler R, Brown T, Rice J (2007) The role of social capital in reducing non-specific psychological distress: the importance of controlling for omitted variable bias. Soc Sci Med 65:842–854

Verhaeghe PP, Pattyn E, Bracke P, Verhaeghe M, Van De Putte B (2012) The association between network social capital and self-rated health: pouring old wine in new bottles? Health Place 18(2):358–365

Index

R. A. Tutu, J. D. Busingye, *Migration, Social Capital, and Health*,
Global Perspectives on Health Geography,
https://doi.org/10.1007/978-3-030-24693-8

The manufacturer's authorised representative in the EU is Springer
Nature Customer Service Centre GmbH, Europaplatz 3, 69115 Heidelberg,
Germany. If you have any concerns regarding our products, please
contact ProductSafety@springernature.com

Printed and bound by CPI Group (UK) Ltd, Croydon, CR0 4YY
30/09/2026
02225465-0015